Space and Society

The Space and Society series explores a broad range of topics in astronomy and the space sciences from the perspectives of the social sciences, humanities, and the arts. As humankind gains an increasingly sophisticated understanding of the structure and evolution of the universe, critical issues arise about the societal implications of this new knowledge. Similarly, as we conduct ever more ambitious missions into space, questions arise about the meaning and significance of our exploration of the solar system and beyond. These and related issues are addressed in books published in this series. Our authors and contributors include scholars from disciplines including but not limited to anthropology, architecture, art, environmental studies, ethics, history, law, literature, philosophy, psychology, religious studies, and sociology. To foster a constructive dialogue between these researchers and the scientists and engineers who seek to understand and explore humankind's cosmic context, the Space and Society series publishes work that is relevant to those engaged in astronomy and the space sciences, while also being of interest to scholars from the author's primary discipline. For example, a book on the anthropology of space exploration in this series benefits individuals and organizations responsible for space missions, while also providing insights of interest to anthropologists. The monographs and edited volumes in the series are academic works that target interdisciplinary professional or scholarly audiences. Space enthusiasts with basic background knowledge will also find works accessible to them.

More information about this series at http://www.springer.com/series/11929

Konrad Szocik
Editor

The Human Factor in a Mission to Mars

An Interdisciplinary Approach

Springer

Editor
Konrad Szocik
Department of Philosophy
and Cognitive Science
University of Information Technology
and Management in Rzeszow
Rzeszów, Poland

ISSN 2199-3882 ISSN 2199-3890 (electronic)
Space and Society
ISBN 978-3-030-02058-3 ISBN 978-3-030-02059-0 (eBook)
https://doi.org/10.1007/978-3-030-02059-0

Library of Congress Control Number: 2019933852

This Springer imprint is published by the registered company Springer Nature Switzerland AG
The registered company address is: Gewerbestrasse 11, 6330 Cham, Switzerland

Preface

Future human missions to Mars will by definition be undertaken by voyagers who are representatives of humanity. The evolved composition of our psychological modules and our cultural histories will constitute the lens through which we experience and interpret these momentous events. Our unique human needs and frailties, and how we plan to satisfy them, will influence both how far we travel and how long we stay away from our terrestrial tethers. Thus, the "human factor" will undoubtedly be the leading consideration for expert and layperson alike as the first Mars missions approach. Several issues are worthy of explanation in regard to "the human factor."

First, the human factor in space missions should be clearly distinguished from uncrewed spaceflights. Humanity has a relatively long history of successful exploration of space by robotically controlled spacecraft. While space exploration by robots has been relatively routine (yet, not so comparable to human-centered exploration when the current robotic capabilities are compared with human dexterity and intelligence), human spaceflights from the past several decades include only routine missions at Earth orbit. Human missions to the Moon realized in 1969–1972 at first glance may look more like irrational extravagancy motivated only by the political competition of the Cold War. But the fact that the human lunar missions were canceled and then never reactivated again is meaningful also in the context of the missions to Mars. It may suggest that no one is able to find a sufficiently strong rationale to once again send human astronauts to the Moon. The canceled human lunar program is not a challenging undertaking in terms of technological challenges with our current state of the art in-flight technology when compared to the challenges of the Mars mission. This lack of a technological challenge undoubtedly casts a shadow upon our collective motivations which are usually spurred by new challenges and difficult enterprises.

Second, the human factor demands a rationale which is much stronger than that for robotic space missions. For many reasons—from financial to medical—human presence in space requires rationale. The lack of technological challenge, political expediency, and other motivators to send humans to the Moon are at least partially responsible for the lack of the human lunar program. There is no doubt that humanity is able to realize large, expensive, risky, and long-term projects on Earth (see: chapter of Jacob Haqq-Misra). For this reason, the question of "go to Mars or not to go" is not purely a financial issue because humanity spends more money for countless terrestrial undertakings (see: chapter of Konrad Szocik). This question becomes an ethical (also in the context of the allocation of financial resources) issue due to the need for a rationale. In some sense, the study of human factor in space is, at least to some extent, identical with the search for a rationale. The readers of this volume may find that almost in every chapter of this book, rationale is discussed to some extent. The last chapter is totally focused on the consideration of this topic. When particular contributors discuss such issues like psychological challenges in space, the impact of space weather, and also ethical issues toward uninhabited Mars, they discuss in fact the rationale for this mission. They show what kind of problems and areas of risks may appear and therefore contribute to the calculus of rationale.

After rationale, the concept of risk is the next key concept connected with the human factor in a mission to Mars. As with the case of rationale, almost every chapter in this volume mentions some risks which should be expected at various stages of the mission. One of the first steps for a successful human mission in space is detection and analysis of all possible and probable areas of risks. This broad set of risks requires an interdisciplinary approach.

Interdisciplinarity is hence the next essential concept. The main idea of this book assumes that the human factor in a mission to Mars will be, in fact, repetition of all problems and challenges which are a domain of human evolution and life on Earth. The extra challenge will be living in a hazardous space environment. This environmental challenge will cause technological and medical problems including unexpected deleterious effects. This environmental factor is expected to affect in a peculiar way the future evolution and development of humans in space.

Interdisciplinarity expresses complexity of the human factor, both on Earth and in space. This case is expressed by areas of expertise of contributors to this volume, and by selected topics. Contributors to this book represent such diverse fields as biomedical engineering, astronomy, astrophysics, physics, meteorology, astrobiology, biology, anthropology, theology, and philosophy. Each of these disciplines expresses a particular facet of being a human being. One could ask if whether it is possible and justified to make some order and hierarchy of priorities and values in these mentioned factors. Another could skeptically reply that technological, medical, and environmental factors should be considered first, or even, that only they matter. Without a life support system and medical protection—which require not only aerospace engineering and medicine, but also many associated disciplines—no one is able to send humans to Mars. But the question arises if skepticism to nonmedical and non-technological considerations may be justified. In our opinion, the answer is no. Human beings include not only physiology and biology, but also culture. Culture mostly in recent years is broadly discussed as a strong selective power comparable to natural selection. Culture is connected with human social and political life. All of them connect with ethical issues. It is hard to imagine and to assume that humans in space will live without culture, and that they will not feel and experience the same needs, feelings, and purposes like they experience on Earth.

For these reasons, the human factor in a mission to Mars as discussed in this book refers to all possible challenges, risks, but also human needs and desires—all contexts of human life as it is known on Earth—which are supposed to appear wherever human factor occurs. It means that humans will take all their specificity and complexity wherever they travel in space. This assumption explains the diverse collection of topics in this volume. However, this list is far from being complete. This book does not discuss such important issues like legal concerns connected with the broad field of space law, or possible future political challenges. This book also does not discuss in detail all the medical challenges which may appear in space (however, a couple of chapters in the first part discuss this issue). One aim of this book is to show how complex and complicated the human factor in a mission to Mars truly is.

This volume is divided into two parts. The first part is focused mostly on medical, psychological, and biological challenges which may appear during human mission to Mars and may be challenging for the success of this mission. This part includes the five following contributions.

Steven Abood discusses the importance of environmental psychology—how our physical environments affect our psychologies, on the Mars mission. Living on Mars will cause numerous psychological challenges. Abood suggests that humanity must attempt to cope with these challenges by the careful design of the choice architecture of environments which make people more or less likely to choose different behaviors. The fulfillment of fundamental existential, relationship, safety, and fitness needs on the Mars mission largely depends on designing environments so that certain choices and behaviors are more favored than others.

Margaret Boone Rappaport and Christopher Corbally study the complexity of challenges caused by the space environment. They focus mostly on the impact of space on human psychology, social life, and culture. Then, they consider a broad set of possible countermeasures to cope with more and less serious threats. These latter ones include, among others, ubiquitous Martian dust, limited access to sunlight, and the monotony of the Martian red color.

Mark Shelhamer is focused on the danger of spaceflights. He shows that even relatively easy and routine tasks conducted in Earth orbit may lead to disaster. Shelhamer underlines the complexity of hazardous and deleterious factors in human spaceflights. He argues that the crew—which should be aware of the fact that the long distance from Earth excludes some rescue and recovery strategies—might be supported by some new tools including, among others, automated systems. However, as the author argues, such solutions may be challenging for our human sense of freedom and autonomy.

The chapter of Mike Hapgood presents the challenges concomitant with space weather. Two the most important factors are here galactic cosmic rays and solar radiation storms. Hapgood reviews their physical specificity and their impact on the health and performance of astronauts. Required countermeasures should include mixed strategies containing both shelter and detecting systems on the board, supported by experts on Earth.

The last contribution in this first part is Chris Impey's chapter. This contribution is a good transition and connection between two parts. Impey connects issues appropriate for the first part—like a broad set of various challenges including even the possibility of new speciation among Martian population, with topics discussed in the second part. These latter ones include such issues like human fascination with Mars, human plans of terraforming Mars, or current, more or less reliable plans of colonizing Mars.

The second part is the collection of nine chapters which explore aspects of the human factors in space other than those comprising technological, psychological, biological, and environmental concerns. The first contribution to this part authored by Klara Anna Capova portrays the idea of Mars colonization from the point of view of sociocultural anthropology. Capova shows how Mars became the planetary pop star which infected human brains in recent centuries.

Jacques Arnould's contribution opens a series of six chapters focused on various ethical questions. Arnould discusses, among others, the politics and policy of space agencies and governments interested in the exploration of Mars. But he also shows why a Mars settlement might constitute the birth of a new ethical horizon for humanity.

Jacob Haqq-Misra focuses his attention on ethical issues on Earth. He notices that the long-term human space program, including a Mars settlement, will require a multi-generational responsible effort aimed at development and maintenance of this mission. Because this effort at least at the beginning will not be profitable for benefactors and sponsors, the author introduces his theory of deep altruism.

Andreas Losch discusses the idea of a human mission to Mars in the broader framework of the concept of sustainability. He argues that it is worth it to include an opportunity to use the outer space both as the part of human heritage, but also as a chance for human survival.

Tony Milligan and Martin Elvis consider one of the possible rationales for a human settlement on Mars: human base for asteroid mining. They point out that it is desirable to apply the 1/8 principle for protection—the final barrier for human exploitation of resources in space which should not extend 1/8 of the available resources to avoid super-exploitation. Their chapter corresponds with Haqq-Misra's care for the long-term responsible maintenance of human space settlement, and with Losch's idea of interplanetary sustainability.

The contribution of Gonzalo Munévar presents the importance of Mars for progress in science. As the author shows, studying Mars matters both for scientific progress in space geology and astrobiology, and also for making progress in science applicable on Earth. He adds that human settlement on Mars should follow purely scientific analysis focused mostly on the study of tracks of life on Mars.

Erik Persson considers possible ethical attitudes toward uninhabited Mars. As he shows in his chapter, the basic—but definitely not obvious and not easy!—task is the necessity to consider the possible intrinsic value of a given object (in this case, Mars). An uninhabited planet does not possess moral status. Persson argues that the main criterion for the application of ethical considerations is existing as a sentient being.

The chapter written by Lluis Oviedo focuses our attention on the issue of religion in space. Oviedo analyzes possible scenarios in the future human Mars settlement in which religious beliefs could play any role. This mental experiment shows how broadly encompassing a human factor is in a mission to Mars. Religion is discussed here mostly due to its enormous role played during human cultural evolution on Earth. Oviedo shows that it is important to consider its possible role in space colony in the context of its psychological and social functions.

The last chapter by Konrad Szocik offers skeptical remarks on the idea of a human mission to Mars. The author enumerates and discusses challenges including the (in his opinion) poor rationale for this mission, the challenge of sustainable development, and the risk for human health and life. While this chapter is without doubts the most skeptical contribution to this volume, it is not aimed at questioning the justification for the human interplanetary mission. Its main idea lies in the assumption that mission planners should take into account many, sometimes not obvious and unexpected, issues. These issues include long-term consequences like the ethical challenge of human enhancement, mentioned in this chapter.

The book shows that the study of the idea and plans for a human mission to Mars is a promising topic of future study through the lens of diverse fields including the social and political sciences. This is a field which is still open for new ideas, especially those which appreciate interdisciplinary, combined efforts. Contributions to this volume show that it is possible and worthwhile to extrapolate facts, processes, and mechanisms known in human history and evolution on Earth to the human future in space.

Rzeszów, Poland Konrad Szocik

Contents

Part I
Environmental, Medical and Psychological Challenges of the Human Factor in a Mission to Mars

Chapter 1
Martian Environmental Psychology: The Choice Architecture of a Mars Mission and Colony

Steven Abood

Abstract The first voyagers who venture to Mars and seek to live on soil beyond our terrestrial home will face an environment mismatched with the one in which their genomes, epigenomes, and psyches evolved. Even if technical hurdles are circumvented to provide adequate resources for basic physiological needs, *Homo sapiens* will not survive on an alien land if a fracturing psychology prohibits the utilization of these resources. Environmental psychology can be employed to shape the choice architecture of a vessel to Mars and a colony upon it, in order to bias choices toward the fulfillment of fundamental existential, relationship, safety, and fitness needs. Aspects of surroundings that should be engineered to create psychological states optimal for survival and welfare include primes, defaults, private spaces, shared spaces, ceiling height, object shape, color, nature, pets, light, windows, noise, temperature, odors, contaminants, order, and diversions. No matter how far we soar into the stars, our psychologies will be ever tethered to the totality of our surroundings. By shaping our environments, we indirectly shape our psyches and prepare them for a mission of unprecedented alienation and duration which might be the last best hope to avert the end of our kind.

1.1 Introduction: Why Concern Ourselves with Martian Environmental Psychology? On Resources and Resourcefulness

In Akira Kurosawa's classic film *Seven Samurai*, a small village of farmers struggle to survive in a hostile land fraught with the exigencies of a country torn asunder by civil war: a shortage of food, fragile shelters threatened by violent weather, and marauding warlords and bandits seeking to take from them the very resources upon which their lives depend. The village elder, Gisaku, proposes that the only way to survive is to hire samurai to protect the village. But one of the farmers, Manzo,

S. Abood (✉)
Department of Biological Sciences, Florida International University,
11220 SW 8th Street, Miami, FL 33199, USA
e-mail: saboo001@fiu.edu

K. Szocik (ed.), *The Human Factor in a Mission to Mars*,
Space and Society, https://doi.org/10.1007/978-3-030-02059-0_1

conveys his worry that the village girls will swoon at the sight of samurai, which will cause problems with the men of the village. To this, Gisaku responds, "Bandits are coming, you fool. What's the use of worrying about your beard when your head's about to be taken?" (Kurosawa 1954).

In much the same way, it may at first appear to be self-evident that focusing on the human factor on a Mars mission and colony is a poor allocation of attention and resources. A critic of human factors may muse that measures to preserve biological integrity protect the head, while measures to enhance psychological welfare preserve merely the beard.

But this is myopic. A human organism that perishes due to a fracturing psychology is just as dead as one that perishes due to a failure in biology. All the resources in the world are useless without the will to utilize them. Too often we fail to survive not because of a lack of resources but due to a lack of resourcefulness. And this lack of resourcefulness emanates from non-optimized psychological states, which make the organism vulnerable to cognitive and emotional storms.

In laboratory experiments, if animals reach a frustrated psychological state in which they believe there is nothing they can do to survive (termed "learned helplessness"), they fail to use the resources available to them to do so (Maier and Seligman 1976; Sweeney et al. 1986). For instance, dogs experiencing learned helplessness will not jump over a small barrier to escape an electric shock. In contrast, dogs not experiencing this psychological state will quickly learn to do so to escape the shocks. A Mars colony will perish if depression and helplessness strips the colonists of their will to survive.

Other psychological phenomena also reveal that any survival calculus must account for more than merely the sum of resources needed to satisfy biological needs. Take suicide. With the exception of the minority of suicides utilized to facilitate an imminent end due to a depletion of biological resources (resources which satisfy the physiological needs of an organism), suicide is a survival failure where physical resources are irrelevant. The World Health Organization (WHO) estimates that there are approximately 788,000 suicides globally each year, and is the second leading cause of death among 15–29 year olds (World Health Organization 2017). In the resource-rich United States, suicide is the third leading cause of death among 15–24 year olds, claiming approximately 44,000 lives each year (National Institute of Mental Health 2017). Suicide has increased by 60% globally over the last 45 years (World Health Organization 2017). The psyche of colonists on a Martian environment will face an alienation from their terrestrial home hithertofore unknown to *Homo sapiens*. As the body's immune system rejects foreign biological matter, the psyche too may reject its foreign surroundings. The existential threat resulting from the possibility of self-harm on a Martian colony should be taken seriously.

Likewise, resources are useless for survival purposes if those seeking to survive are too busy fighting one another to use them or use the resources themselves to destroy one another. The story of the Darwinian success of modern *Homo sapiens* is largely intertwined with the abundant natural resources on Earth that were optimized by technological innovations made possible by our enhanced cognitive capabilities. For instance, during World War I, with a rapidly expanding world population on the brink of starvation, two chemists, Fritz Haber and Carl Bosch, discovered how to pull nitrogen from the air and fix this nitrogen in fertilizer that could be used to support more crops than ever before. Their process, aptly named the Haber–Bosch process, increased the capacity of the world's food supply and helped support a population explosion where in a little over a century, *Homo sapiens* multiplied from 1.6 billion people in 1900 to over 7.4 billion today. The amount of nitrogen extracted from the atmosphere by the Haber–Bosch process applied to crops during the decade from 1980 to 1990 is greater than that from all industrial fertilizer applied previously in human history. It now produces more than 450 million tons of nitrogen fertilizer per year, and nearly 80% of the nitrogen found in human tissues originated from the Haber–Bosch process. In combination with pesticides, the Haber–Bosch process has quadrupled the productivity of agricultural land (Smil 1999; Erisman et al. 2008). However, all these resources and others have little utility in satisfying fundamental survival needs if coercive political regimes engender conflicts that necessitate the diversion of resources toward military spending. The resources themselves vanish if these regimes cause conflicts which destroy resources. The 5.8 trillion dollars of spending on nuclear weapons by the United States in the Cold War to attempt to avert the spread of Communism and the burning of the Kuwaiti oil wells by Saddam Hussein are two cases in point (Wald 1998; Salgado 2016). The presence of significant interpersonal conflicts on simulated Mars missions and colonies, as well as in analogous survival situations, may herald conflicts of even greater intensity on an actual Mars mission and colony.

These psychological phenomena, as well as others, result in survival failures despite resources adequate to satisfy physiological needs. No matter how adequate the resources on a Mars mission and colony, then, the human factor can frustrate survival. It is the fly in the technological ointment.

On a Mars mission and colony, the unprecedented distance from Earth and duration of the mission will magnify the psychological challenges for the voyagers involved. The burden of establishing a multi-planetary home to potentially avert the extinction of our species and the accumulation of all our collective knowledge and memories in the case of a mass catastrophic event (such as the one 252 million years ago which destroyed 90% of life on Earth, or the one 66 million years ago which decimated 50% of life) (Benton 2003), will fall heavily upon the shoulders of these men and women.

How will our terrestrial psychology face these challenges amidst the stars and on Mars? We can draw analogies from polar expeditions and other survival situations and see that it may be necessary to sometimes supplant the whims of the individual in favor of the survival needs of the group (Szocik et al. 2018; Kansas 2010). Although much has been said about the selfish gene, even cooperation entailing self-sacrifice may ultimately serve selfish ends through mechanisms such as kin selection (Wilson 1975). Evolutionary drivers of behavior such as this are difficult to oppose. How can we augment this cooperation and other sociological and psychological attributes that will encourage mission success through an engineered environment? And as we strive for cooperation to facilitate survival, how can we encourage cooperation in the least coercive manner possible? Much of our recent enlightenment as a species has been the onus we place on the freedom of the individual. Can we encourage cooperation by suggestion rather than coercion? It would seem to be one step forward for humankind and two steps backward if we established a Mars colony on the foundation of despotism. As we balance the existential need for our species to live, with our preference to live well, that is, to live freely, environmental psychology can play a role in maintaining freedom to choose while increasing the likelihood that those choices will serve both the individual and the group.

1.2 Solving the Resourcefulness Problem: A Brief Introduction to Environmental Psychology and Choice Architecture

Environmental psychology refers to how our surroundings affect our psyches and behavior (Ittelson et al. 1974). Physical structures can enable and influence the likelihood of thoughts, moods, and behaviors. Anyone who had stood within the awe-inspiring cathedrals of Europe, or stood before the Great Buddha of Thailand, can appreciate how architecture can elevate the psyche. In contrast, anyone who has spent time living in the cramped quarters of an inner city dwelling can understand how architecture can also discourage and depress. We exist not as brains in a vacuum but as nodes in an inextricably interconnected system of connections and inputs which affect us. Like cells whose responses can only be truly described in the context of their extracellular environments, we can only truly describe our own thoughts, moods, and behavior in the context of the other minds and things which surround us.

Choice architecture is a related concept and refers to the design of how different choices are presented to us which can make it more or less likely for us to choose a certain path (Scheibehenne et al. 2010). Sometimes, the arrangement of these choices can actually be embedded within the physical architecture of our surroundings, and this is where environmental psychology and choice architecture converge. For instance, placing your refrigerator and pantry in your bedroom, as opposed to placing your refrigerator and pantry far away from your bedroom—across the house inside your garage—will strongly influence your decision whether to engage in late night snacking or not.

The design of our surroundings can influence our choices in numerous subtle ways. Simply etching the image of a black housefly into a urinal causes bathroom patrons to hit their target 80% more often (Pollack 2012). Beyond this crude example, simple architectural choice can influence countless behaviors through various means.

This has led some to postulate that there is no such thing as a neutral architectural design. Design affects behavior. In perhaps an overextension of these ideas, the theory of architectural determinism proposes that the surrounding environment is the chief or even the sole determinant of behavior (Broady 1966).

It may be more accurate and less authoritarian, to conceptualize our surroundings as retaining the potential to strongly determine our thoughts, mood, and behavior, as opposed to absolutely determining them. This conception has the advantage of construing design efforts as liberally paternalistic, as opposed to coercive, which is more compatible with our onus to preserve freedom of choice.

On a Mars mission and colony, compliance with certain directives that are essential with facilitating survival will be imperative. For the survival of the voyagers who will become colonists, this is essential. If we view these colonists as the possible last bastion of humanity upon which the survival of our species depends, or if catastrophic events on Earth makes this so, the import of compliance with survival directives takes on an entirely new meaning. In this case, a different calculus, one in which more coercion is at least temporarily necessary as the colony reaches a critical self-sustaining mass, may be appropriate.

Still, we often catch more flies with honey than with vinegar. Choice architecture can help bias colonists toward compliance regardless of the presence or absence of strict behavioral rules.

The environmental design may seek to bias the individual toward a specific behavior. For instance, the design of living quarters and private and public space on a spacecraft to Mars may attempt to bias individuals to interact with all the other shipmates as opposed to only a select subgroup of them. Since we know that social rejection detrimentally affects mood and the immune system (Murphy et al. 2012), encouraging social interaction and social acceptance in this way can have important beneficent effects.

In addition, the environmental design may seek to bias the individual toward a particular mood that will facilitate a number of nonspecific, but proactive behaviors. Moods which facilitate a positive explanatory style have been found to facilitate goal achievement in a variety of venues (Seligman 2006). Positivity increases resourcefulness regardless of resources.

An early application of environmental psychology concepts was the Panopticon, a design for a prison facility described by Jeremy Bentham in a 1798 essay (Bentham 1798). The goal of the Panopticon was to engender in the inmates a feeling that the authorities were virtually omniscient, and could be watching the prisoners at any moment. To achieve this, the Panopticon was designed as a circular building with the prisoners in the circumference and the guards in the center. The design maximizes visibility of the prisoners by the guards but through various architectural contrivances, largely shields the guards from the view of the prisoners (Foucault 1979; Johnston 2000; Semple 1993). Only a few Panopticons were built, but a recent design, the Federal Metropolitan Correctional Center (MCC) in Chicago, seemed to draw upon its concepts. The Chicago MCC was shaped like an equilateral triangle with the guard station in the middle. Almost all spaces outside of sleeping quarters could be viewed from the guard station. In addition, the shape gave all rooms an outside view while providing for an efficient use of interior space (Werner 2012).

Environmental psychology and choice architecture was integral to the computer game SimCity, which was inspired by Christopher Alexander's 1977 book on architecture and urban design entitled *A Pattern Language: Towns, Buildings, Construction* (Alexander et al. 1977). SimCity's creator, Will Wright, was intrigued by the book's focus on functionality over aesthetics in architecture, and wanted to create a game that would encourage certain human behaviors through architectural design. Wright, who had lost his life's possessions in a fire, sought to create a miniature virtual world where simulating life itself was the object of the game (Whitehead 2008). SimCity was the first world building game, and each player is given the task of designing his or her own city from a patch of green land. All the decisions about where to place residential zones, commercial zones, and industrial zones affect the decisions the sims (the miniature inhabitants of the city) make and their happiness. The sims live their lives and make their choices—where to live, who to marry, whether to have babies or not, subconsciously influenced by the choices the game player makes as the grand designer of their world and the architecture contained within it.

The architecture of the Mars mission and colony, like the choice architecture in the sims game, will have profound effects on behavior. Psychological leverage can emanate through careful environmental design resulting in an architectural jiujitsu that encourages certain beneficial states of mind and behaviors that facilitate survival. These designs may manifest in smart architecture which change shape or display based on the preferred behavior that needs to be encouraged in different situations.

Where should the architects of humanity's future on Mars begin? It would be prudent to start with a few essential aspects of choice architectural decision which we know affect behavior on Earth and apply these to the needs of the Mars mission and colony when creating architectural designs for them. The following discussion surveys a few of these fundamentals.

1.3 Primes

Priming influences behavior by exposing a person to one stimulus, which then biases his or her response to subsequent stimuli.

In one experiment demonstrating how powerful priming can be, students were given a task where they had to create four word sentences from lists of five words. For example, the five words "from are Florida oranges temperature," would be rearranged to form the sentence "Oranges are from Florida." After completing the task for lists of five words, the students left the room. The time it took for them to walk down the hall was secretly measured. It turned out that students who had been given words associated with old age such as "Florida", "forgetful", "bald", "gray", or "wrinkle", walked down the hall much more slowly than students who did not rearrange such words. Although none of the students reported noticing that the words had a common theme, they were subconsciously influenced to move in a way associated with old age (Bargh, 1996).

These subconscious effects can also be used to increase goal-related performance and grit. In another experiment, twice as many participants who rearranged words related to success and goal achievement continued to work on a puzzle after being told to stop, than those who did not rearrange success-related words (Bargh et al. 2001).

Primes, like other aspects of our surroundings that bias behavior, often work subconsciously. The majority of our mental processing exists below the level of our conscious awareness (Koch 2009; Zimmerman 1989). A key entry point for subconscious information is the eyes. The evolutionary needs of hunting and gathering, as well as fleeing or fighting predators, ensured that the human eye channels information, both consciously and subconsciously to the brain. In a phenomenon called blindsight, blind individuals can identify whether a picture depicts a happy or angry face at levels better than chance (Toro et al. 2008; Pegna et al. 2005; Ekman and Friesen 1975). While their conscious visual pathways are cut off, their subconscious visual pathways are still intact. Visual displays encouraging positivity, resilience, and cooperation will be particularly important on the Mars mission.

Visual primes and primes utilizing other modalities on the Mars mission and colony could encourage behavior to enhance health and wellness in four fundamental areas of human needs. These fundamental human needs—drawn and slightly modified from the psychologist Abraham Maslow's hierarchy of needs (Maslow 1943)—include physiological, security, relationship, and existential needs. The optimization of measures to satisfy each of these needs is particularly important in the atmosphere of the Mars mission and colony which represents a genetic, epigenetic, and psychological mismatch to the terrestrial environment in which humans evolved.

The first of these, physiological needs, relates to all the biological needs of the organism including energy intake (adequate food and nutrition), energy conservation and restoration (rest and sleep), and maintenance and growth (through stressing the body through exercise, for instance) (Williams et al. 2009). Primes that encourage exercise measures that are designed through choice architectural methods for optimal convenience and interest could facilitate the satisfaction of this need. For instance, an augmented reality game that an occupant is visually reminded is accessible in the private quarters, or communally played in the shared quarters that solves a mystery or engages the participants on some adventure together that requires movement, may bias participants toward regular exercise. In 2016, a terrestrial augmented reality game, Pokemon Go, increased the activity level of participants by 1473 steps a day on average, a more than 25% increase. In only 30 days, the game added a total of 144 billion steps to U.S. physical activity levels (Althoff 2016). Exercise involving the learning of new complex movements involving bilateral coordinative functions, such as dance, martial arts, sports, and yoga, could not only improve compliance through increased enjoyment but also enhance cognitive function and mood (Budde et al. 2008). Mechanistically, this might occur when exercise produces the hormone atrial natriuretic peptide (ANP). When heart rate increases during exercise, more ANP is produced which counteracts stress hormones, resulting in a decrease in stress and anxiety (Koopmann et al. 2014). Additionally, exercise boosts the production of natural opiates called endorphins, which fight pain and boost mood (Daniel et al. 1992). Not only does exercise boost ANP and endorphins, it raises levels of the neurotransmitters targeted by antidepressant medications, including dopamine, norepinephrine, and serotonin (Lin and Kuo 2013). All three of these monoamines increase mood and are part of the mechanism of action of various antidepressant medications. Exercise also boosts brain-derived neurotrophic factor

(BDNF) (Sleiman et al. 2016). Chronic stress raises cortisol levels, which cause neurons and their connections to shrink in the hippocampus, a cluster of brain cells essential for learning and memory (Zhang et al. 2016). BDNF, conversely, causes connections between neurons to sprout and grow, creating the neural infrastructure through which the monoamines can exert their effects (Sleiman et al. 2016).

Next in the hierarchy of needs are those related to safety or security, which encompass all the concerns that humans have about remaining secure from the perils of their environment, whether such dangers come from natural or created hazards, the malevolence or dominant hierarchy-reinforcing intentions of other humans, the nutrient needs of other organisms, or the self-propagating directives of microorganisms, viruses, or prions. Primes in this category could remind and encourage colonists to take adequate security and safety measures.

The third category of fundamental human needs is the area of relationship needs, the satisfaction of our innate desire for love and belonging. Primes that encourage patience, cooperation, and empathy could facilitate the satisfaction of these relationship needs.

Last in the hierarchy are needs related to existential fulfillment, which might be satisfied through adherence to various goal-achievement training methodologies. Our *telos*, or purpose, can be the essential driver to stave off debilitating conditions such as anxiety and depression. A key method of satisfying this need on a Mars mission could be psychological diagnostics beforehand which identify a number of subgoals (apart from the overall mission) which give the participants' lives meaning, and the facilitation of goal-achievement methods to engage in the pursuit of these subgoals on the mission such as specifically tailored primes. These could include messages on physical displays projected onto walls of private living quarters, displays embedded in biomechanical enhancements, wearable technology, virtual reality or in augmented reality, which help maintain a state of interest, positivity, and enthusiasm about subgoal achievement.

1.4 Defaults

We tend to make decisions based on certain cognitive rules of thumb or heuristics (Tversky and Kahneman 1974). The environment can be shaped to bias decision-making especially when the shaping is in accordance with these heuristics. One heuristic is the status quo bias. This is the tendency for us to accept the default position, to succumb to the inertia of our circumstances and surroundings (Samuelson and Zeckhauser 1988).

On the Mars mission and colony, the more choices that can be engineered into the environment as the default position to conserve resources and promote the health and welfare of the entire community, the more likely that compliance that facilitates survival will be achieved. And the resulting compliance from defaults will then be the result of a liberal paternalism as opposed to a more coercive compliance. An example of coercive compliance would be a medical device such as an IntelliDrug implant, a prosthetic tooth that automatically releases medicine into a patient's mouth (Moscicka et al. 2007).

Since the default setting, i.e., what choice comes first, must be decided upon anyways, and if architecture signals us to act in one way or another, defaults can be used to benefit the group by strengthening individual health, encouraging consideration for others, and conserving resources. For instance, in public bathrooms in the United States, water spigots are often programmed to default to shutting off after a few seconds of running, to conserve water. Defaults can be used to maintain freedom of choice—a person can still push the button and receive more water if she wants, after the spigot had been programmed to stop, while biasing choice in one direction. On a Mars mission and colony, this biasing will be toward decisions that promulgate the survival and welfare of the community.

The status quo bias may partly exist because of another heuristic that makes us loss averse. To simplify what the loss aversion heuristic is all about, imagine having five dollars and then losing it. Now imagine a different scenario where you gained five dollars. You would be much more upset if you lost five dollars than you would be happy in gaining five dollars. Losing something you already have makes you (approximately) twice as upset as gaining that thing would make you happy. In the case of the status quo bias, the thing lost is the default option that you started with. In one experiment on loss aversion, some students were given mugs and some were not. Those with mugs demanded roughly twice as much money to part with their mug, as the students without the mugs were willing to pay to get one (Kahneman et al. 1990). The choice architecture of the Mars colony and mission can be biased by acting in accordance with these and other heuristics toward choices that benefit the survival and welfare of the entire community.

1.5 Private Spaces

The choice architecture of private spaces can either serve to replenish energy reserves or help engender interpersonal conflicts. Let us examine private spaces in terms of three components: privacy, personal space, and territoriality.

1.5.1 Privacy

Privacy refers to the ability to seclude oneself from others. The degree of privacy that is preferred varies from individual to individual and within the same individual in response to different surroundings and at different times (Altman 1981). Being alone can be a positive or negative experience based on preference and represents the difference between solitude—preferred aloneness—and isolation—non-preferred aloneness. Privacy can be an important avenue for an individual to rejuvenate from stress, contemplate emotional issues, process different circumstances, and exert a sense of control over proximate space that can be a virtual extension of the self.

Living quarters where privacy is not an option can be a significant source of strain (Evans 1982; Evans and Stecker 2004). Crowding occurs when a person seeks privacy but is denied due to her lack of choice to restrict individuals away from her.

The architecture of living quarters on the Mars mission and colony should be optimized to ensure the choice of privacy when desired.

1.5.2 Personal Space

Density, how many people there are per unit of physical space, does not necessarily cause crowding, which depends on a person's psychological preference to be away from a certain number of people or particular people. As would be expected, individuals with more aggressive personalities feel more crowded at closer distances than others and require larger interpersonal distances (Kinzel 1970; Curran et al. 1978; Hildreth et al. 1971; Roger and Schalekamp 1976). An invasion of personal space is a significant cause of aggression (O'Neal et al. 1979; Ryden et al. 1991; Bridges-Parlet et al. 1994; Fagan-Pryor et al. 2003; Daffern et al. 2004), and having personal space reduces the likelihood of aggression in correctional settings (Nijman et al. 1999). Correctional settings can serve as useful model systems to draw lessons for planning the journey to Mars, since the relative permanence of the confinement, lack of ability to leave, isolation from friends and family, and confinement with strangers, comprise similar aspects, that are not all present in many laboratory settings or even simulations of a Mars colony.

Social density refers to the number of individuals per unit of space, while spatial density refers to the amount of personal space a person has to themselves (Braum and Paulus 1987). Increasing the total number of people in a room would be increasing the social density, while keeping the spatial density constant. Research in laboratory and real-world settings has generally shown that although high densities of both kinds can increase stress, increasing social density is most likely to cause aggression and interpersonal conflicts (Braum and Paulus 1987). On a Mars mission and colony, it would be therefore best to partition a space into many smaller rooms that are occupied by less people, rather than having more people in larger rooms.

Shared sleeping spaces may violate primordial concerns about safety and security while sleeping. In correctional institutions, where safety concerns will obviously be magnified, inmates experienced heightened markers of the stress response such as elevated blood pressure when moved to shared sleeping quarters, which subsided when they were moved back to private spaces (D'Atri and Ostfield 1975). The perception of increased crowding is also related in inmates to a greater perception of the existence of hostile intentions in others (Ray et al. 1982; Roush 1989; Lawrence and Andrews 2004).

So the preference to avoid social density and obtain a private space, even a small one, may stem from safety and security needs, a desire to avoid an overload of stimuli bombarding the senses (Milgram 1970), or a desire to avoid unwanted and unpredictable social interactions that may block the obtainment of goals and preferences (Langer and Saegert 1977).

The number of resources per unit of space can exacerbate stress, since scarcity leads to competition, which increases the likelihood of aggression (Mequida and Wiener 1996).

1.5.3 Territoriality

Territoriality refers to how and to what degree people assert ownership and dominance over a geographical area and the specific objects and structures within it. Territorial behavior may take various forms such as occupying a geographical space, defending a space or attacking the occupants of a space, moving personal objects into a space, or marking a space or objects within it.

The need to assert dominance over a territory is ubiquitous throughout the animal kingdom and is deeply rooted in our biology. In a massive study of 18 chimpanzee communities for a period of 426 researcher years, 63% of chimpanzees killed by other chimpanzees were attacked by chimps from outside their own in-group (Wilson et al. 2014). This percentage does not account for additional violence and killings over territorial disputes within in-groups. Disputes over territory account for a significant amount of human aggression and violence (Paluck and Esser 1971; Ardrey 1967; Van den Berghe 1974).

There is less interpersonal aggression and violence from territorial disputes when dominance hierarchies are stable and the population is relatively fixed. In studies of juvenile detention centers, those at the top of dominance hierarchies had the choices of what territories they wanted to occupy (Sundstrom and Altman 1972). When the population was stable, i.e., when there was not a significant influx or efflux of boys into or out of the facility, the same boys occupied the same geographic spaces with considerable consistency. Aggression and interpersonal violence was low. However, when there was considerable turnover in the population, the hierarchy was in flux, and there was little consistency and much uncertainty over who had the power to occupy what spaces. Interpersonal violence over territoriality in these times of population flux was high. Once a new hierarchy was established, and territories were occupied with consistency, interpersonal violence again decreased.

To create a viable Mars colony, more and more colonists will arrive and continue to settle as communities attempt to achieve a self-sustaining critical mass. If issues of territoriality are not clearly demarcated, the potential for interpersonal aggression will be significant, especially when these colonists will likely come from a diversity of Earth nations and cultures.

On a Mars mission and colony, the earlier territories are marked, including the territorial issues in shared personal quarters, the less potential for interpersonal aggression. The choice architecture of personal quarters should be set at a default so that property and space are clearly demarcated as to whom it "belongs" to. This will be especially important since the duration of isolation is positively correlated with an increase in territorial assertions. In a study of Navy vessels where two individuals were isolated together for a long period of time in a shared private space, territorial behaviors increased over time (Altman and Haythorn 1967). If territorial boundaries were set early on (who had the right to use which drawers, what amount of closet space, what personal objects, etc.), interpersonal conflict over territory was minimized.

On a Mars mission and colony, clearly demarcating some areas as public may minimize conflicts and encourage social interaction. In a study of territorial behavior in three Canadian correctional facilities, the most frequent and positive social interactions were found in clearly delineated public spaces that were not "owned" by any one group or individual (Cooley et al. 1973). Separating territories into primary territories (such as aspects of living quarters that "belong" to one person or the other), secondary territories (workplaces where a person has some territorial rights, but so do others), and public territories (shared public spaces to which everyone has a right to) (Brown 1987) may decrease the prevalence of interpersonal conflict between Mars voyagers.

Low-cost and largely symbolic signs of separation marking territorial boundaries can also decrease interpersonal conflicts on the Mars mission. In humans and non-human animals, demarcation of boundaries is often signaled by territorial markings or symbolic barriers as opposed to constituting impermeable physical boundaries (Brown 1987). The mere addition of a low partition around an otherwise open bunk in a dormitory room where many inmates slept decreased perceptions of crowding and stress in a dormitory to levels near those of single cells (Paulus 1988). Single bunks instead of double bunks (beds stacked on top of one another) and cubicles also decreased perceptions of crowding (Paulus 1988). For the Mars mission and colony, perhaps architectural analogies to soundproofed Japanese capsule hotels, where numerous people can be stacked into capsules that are completely separated physically and visually from one another, will be beneficial to preserve territorial needs despite space constraints.

1.6 Shared Spaces

As much as people have a need for solitude to separate from the cacophony of other people, they also have a need to connect and socialize with others. The psychological pain of isolation can be so profound that it is used as punishment in prisons in the form of solitary confinement (Grassian and Friedman 1986; Haney 2003; Liederman 1962). Solitary confinement can be particularly damaging if sensory input is deprived. Even short-term sensory deprivation of a few hours can lead to anxiety, agitation, and hallucinations (Bexton et al. 1954; Liederman 1962; Lilly 1956; Zubek 1969).

In a study of 6,500 individuals, it was found that prolonged social isolation impairs health and leads to increased mortality independent of loneliness (Cole 2015; Steptoe et al. 2013).

We are a tribal species. Through eons of our history, *Homo sapiens* and our ancestors bonded together for protection and social connection. It is thought that the feedback of others is integral to our understanding of self, and our surroundings (Burke 1996). Therefore, too little social and sensory input, just like too much in the case of crowding, can be deleterious (Liederman 1962).

We also turn to others to confirm our interpretations of sensory input. When we are alone or in a small group, we naturally feel more physically vulnerable and have heightened startle reflexes (Poli and Angrilli 2015). As studies of isolated arctic expeditions reveal, the combination of lack of companions to confirm interpretations of sensory input, a baseline anxiety due to feelings of physical vulnerability due to isolation, and possible auditory and visual hallucinations due to sensory deprivation, can lead to fear and panic (Barnes and Gibson 2013; Harrison et al. 1991). Real and imagined sensory messages can be given the most malevolent interpretations in the fog of such psychological states. One study suggests that when we encounter stimuli we are evolutionarily unprepared to process, or when surroundings are monotonous, due to a lack of social interaction, for example, our left temporoparietal junction can create hallucinations of an illusory shadow person that mirrors changes in the Hallucinator's body position and posture (Arzy et al. 2006). These psychological effects may be exacerbated by the monotony of the Martian landscape (Cockell 2002). On the foreign environment of the Mars surface, exploratory missions should be comprised of large groups whenever possible, to avoid these serious issues.

The more the individuals around you, the more the potential for unexpected and novel occurrences, communications, and interactions. Boredom, therefore, is often associated with isolation. And boredom is far from innocuous. Numerous studies show that boredom is linked to both depression and hostility (Vodanovich 2003). Festivals, holidays, parties, and other events will be important on the Mars mission and colony, to decrease monotony and to encourage the congregation of people whose lack of proximity in living quarters may otherwise fail to bring them together.

Increased stress often leads people to seek the social support of others (Burke 1996). People with a strong network of social support are much less vulnerable to the negative effects of stressors (Frone et al. 1991). In contrast to social support which ameliorates psychological strain, social rejection leads to significant psychological distress (Williams and Nida 2011). During the Biosphere 2 simulation, in which participants lived in a greenhouse-like habitat to test the viability of closed ecological systems to support and maintain human life in outer space, two crew members were so distressed by one another than they did not speak to each other for 18 months, apart from mission-critical exchanges (Nelson et al. 2015). Because of the alienation from friends and family at home, and the difficulties of conversing with them even electronically, choice architecture which encourages social support networks will be particularly important. A psychologist for each vessel would be ideal. An artificial intelligence serving therapeutic needs could serve as a substitute for a human presence. Pre-recorded electronic messages from family and friends offering words of encouragement and advice for different hardships may also serve as a social support adjunct.

To avoid many of these issues, it would be prudent to design shared spaces so that as many different people as possible are likely to interact with one another. Such a utilitarian design would lead to a greater plurality and novelty of interactions, decreasing the problems of lack of social interaction and sensory input described above. To realize such a design, it is important to recognize that social closeness is directly correlated with proximity. In one study, despite all the reasons one could have to establish a friendship, 90% of cadets in a military academy named the person they initially sat next to as the person they had formed the closest friendship with (Segal 1974). When it comes to relationships, proximity is destiny. In a study of college friendships, students who were assigned to live at the ends of the housing complex, where there was the least foot traffic, became relative social outcasts. Those living at the center of the housing complex, where there was the most foot traffic, became the most popular students, with the most social connections (Festinger et al. 1950). Another study showed that scientists who worked in different labs on the same floor were six times more likely to form professional relationships to collaborate with one another, than scientists on separate floors of the same building (Kraut et al. 1990).

In light of these social bonding effects of proximity, the Panopticon design, originating from the need to keep people away from society, perhaps could be employed to bring people on the Mars mission together. A circular layer-caked design of living quarters stacked on top of one another in circular rings that all open up to a center could be utilized for space vessels to Mars, and perhaps even above ground and below ground habitats on Mars. The differences from the Panopticon of old would be that the center would be the shared space (instead of the guard station) and instead of openly visible jail cells, the private quarters circularly surrounding the center, would be closed off. In this way, the center occupants would not be able to view the surrounding circle as they would in the original Panopticon design. And the inhabitants would be able to seek the solitude of their private living quarters when they so chose. But the circular architecture of the design would confer the popularity and social benefits of proximity on everyone, since everyone would be more or less equally likely to mix with everyone else in the shared space in a kind of Arthurian egalitarianism. Like the knights of the round table, no person would be isolated at the ends. And because of the circular design devoid of interior private sleeping quarters, each sleeping quarter would enjoy an outward view, necessary for the orientation of time and place.

Within the shared space, further architectural psychology measures could be utilized to bias the occupants toward social interaction. The more comfortable the furniture to lie and sit, the more likely people will spend more time there, instead of moving back into their private quarters. The arrangement of furniture and other objects in the space can encourage traffic to back up and become crowded at certain places within it like objects directed through a funnel. Perhaps, these places could be multiple centers of gaming activities, as well as other centers where people seeking social interaction but less stimulation can simply talk. The direction the furniture encourages people to face is more influential than the physical distance between them in facilitating interaction. Furniture spaced close together but which do not position their occupants face to face with one another, such as seats at the counter of a diner all facing the kitchen, confer a sense of privacy despite their proximity. In contrast, seats facing one another encourage the initiation of social interactions. In addition to these features of composition, arrangement, and direction of furniture, removing signals of time, a tactic that casinos often engage in, often encourage focus on present activities, instead of other goals and preferences which may encourage individuals to leave the place in which they find themselves (Knapp et al. 2014).

Music, and the dance that often accompanies it, can also encourage congregation (D'Ausilio et al. 2015). However, music in shared spaces should be played during limited times, so as not to disturb those who prefer some sociality but not the kind or presence of music. Wearable technology could also be employed to arrange it so that those who wish to hear the same music or listen to their own music but not disturb others are able to do so, and those that wish to be around other people but be without music, don't.

1.7 Ceiling Height

Ceiling height may affect thinking style. In one experiment, people in rooms with lofty ceilings engaged in more abstract forms of thinking, able to expand their viewpoint and engage in more relational processing between different items and concepts (Meyers-Levy and Rui 2007). In contrast, people in rooms with low ceilings engaged in more item-specific processing, where they directed their attention on a single focal point. As both kinds of thinking are necessary for different kinds of challenges, it may be useful through smart architecture projections upon interior walls of private living spaces, or through virtual or augmented reality, to change the appearance of the perceived vastness of space depending on the kind of challenges the individual is faced with during the Mars mission and colonization.

1.8 Object Shape

Interior furniture with curvilinear forms results in significantly higher feelings of relaxation, peace, and calm, than furniture with rectilinear forms (Dazkir and Read 2011). Given the importance of adequate sleep and the endocrine dangers of circadian disruption, curvilinear furniture in private spaces where colonists will sleep may be particularly important.

1.9 Color

The color of the surroundings may help shape mood. A global survey of art preferences found that blue, followed by green, was the preferred color (Elliot and Maier 2003; Dutton 2003). Our preference for these colors may have evolved because of the survival advantage that a preference for a bright blue sky, blue water or lake, or a green landscape, may have conferred in hunting, gathering, and defending from predators and other enemies. For example, a bright blue sky may have encouraged early *Homo sapiens* to venture outside, explore, and gather food, and a preference for viewing a wide blue sky may have favored those who could build open settlements where the sky, as well as predators and enemies, could be seen from a distance. A wide open landscape of water or vegetation may serve as a natural barrier against predators and enemies. Because of the apparent calming effects of the color blue, some cities in Poland and Japan have installed blue street lighting in an attempt to decrease crime and suicides (Shimbun 2008).

Modes of cognitive problem-solving may also be influenced by color. While red is often associated with danger and make people more alert or aware, blue is associated with increased imagination and creativity. In a study where psychologists had participants perform cognitive tasks with geometric shapes, participants generated twice as many creative solutions when immersed in blue surroundings than when surrounded by red (Mehta and Zhu 2009). In the same study, participants did better on tasks requiring short-term memory when surrounded by red than by blue, perhaps highlighting the color red's role in focusing attention on the present moment, as would be appropriate when responding to imminent danger.

1.10 Nature

The theory that we prefer specific colors in different situations (discussed in the previous section) to enhance survival is a part of evolutionary aesthetics, which refers to evolutionary psychology theories in which the basic aesthetic preferences of *Homo sapiens* have evolved in order to enhance survival and reproductive success. Biophilia is an evolutionary aesthetic theory proposed by E.O. Wilson (Wilson 1984) to describe the innate tendency of humans to seek out nature and other forms of life, and the effect that this tendency has on their psychology and biology. The "savanna hypothesis," proposed by Gordon Orians, states that we are hardwired via evolution to experience beneficial biological and psychological states in the presence of beneficial primordial habitats (Lohr 2007).

Various studies show that natural surroundings help lower stress by affecting emotions. One study demonstrated that nature scenes significantly decrease fear arousal (Ulrich 1979). A study that measured alpha brain waves (Electroencephalography, EEG-a/EEG-b) and forehead electromyography (EMG) showed decreased anxiety in natural settings (Chang et al. 2008). Another using EEG and EMG showed that activities with plants (especially flowering plants) promote physiological relaxation (Yamane et al. 2004). Nature views were found in experiments with adolescents and adults to help facilitate a restoration of calm after experiencing a threatening or emotionally negative event (Korpela 1992; Korpela and Hartig 1996; Korpela et al. 2001). A view of nature through office windows helped decrease job stress when compared to those who did not enjoy a natural view (Leather et al. 1998). Likewise, inmates with external views of nature have reduced blood pressure when compared with those who do not (Moore 1985). And gardens have been found to effectively reduce the emotional distress of hospitalized children (Sherman et al. 2005). Some of these stress-reducing effects may be partly due to reductions in aggression and anger when viewing natural scenery such as while working in gardens (Mooney and Nicell 1992).

But with import for the Mars mission, the stress-reducing psychological benefits of nature need not derive from views of actual nature. Depictions of nature on office walls decrease stress and reduce anger (Kweon et al. 2008). Other studies also show decreased stress, blood pressure, and heart rate from views of nature scenes in art, photographs, and film (Nanden et al. 2008; Ulrich 1999).

These stress-reducing effects may be responsible for the enhanced immune function of those immersed in natural settings. An analysis of a decade of stays at one major hospital showed that surgical patients assigned to rooms with windows overlooking natural scenery had shorter postoperative hospital stays and took fewer potent analgesics than patients in similar rooms with windows facing a brick wall (Ulrich 1984, 1999). In another study, appendectomy patients in hospital rooms with plants and flowers had significantly less intake of postoperative analgesics, more positive physiological responses evidenced by lower systolic blood pressure and heart rate, and lower ratings of pain, anxiety, and fatigue (Park and Mattson 2008). Natural Killer Cells (lymphocytes which play an integral role in the innate immune system) and intracellular anticancer proteins are elevated for more than 7 days in city-dwellers who took a day trip to the forest (Li et al. 2008). Nature scenes combined with nature sounds reduced pain in patients undergoing flexible bronchoscopy (Diette et al. 2003).

Exposure to natural scenery is also associated with improved cognitive performance and attention. Small amounts of green space in housing projects were found to be associated with improved attention span (Kuo and Sullivan 2001). In another experiment, nature exposure increased performance on mentally challenging tasks (Hartig et al. 2003). NASA has contemplated the effects of exposure to natural scenery to counteract both sensory deprivation and social isolation (Bachman et al. 2012).

1.11 Pets

Like natural scenery, pets have also been part of humanity's evolutionary past. Recent archeological findings show that dogs have been buried alongside humans, complete with items such as decorative collars and spoons which they would utilize in this afterlife (Losey et al. 2011). This evidence points to a close relationship with our canine companions.

On a Mars mission and colony, since resource allocation will prioritize humans over pets, artificial pets may serve as a substitute to conserve resources. An embodied artificial intelligence in a pet's body may be useful as long as adequate safety measures are put in place. Handheld digital pets called Tamagotchi, developed by Akikiro Yokoi, were popular in Japan in the late 1990s and 2000s (Wallace 2018), and artificial intelligence has made significant strides since then. Since fMRI studies show that touch dampens the stress response, and other studies demonstrate that it increases levels of oxytocin and decreases blood pressure (Coan et al. 2006; Light et al. 2004), embodied pets would probably be preferable to a virtual or augmented reality pet. In one study, patients recovering from heart disease who had pets had greater 1-year survival rates than those who did not (Friedmann et al. 1980). Immunoglobulin A (IgA) levels were boosted in participants who petted a dog, while those who petted a stuffed toy dog, or simply relaxed on a couch, did not experience significantly increased IgA (Charnetski et al. 2004, Nagasawa et al. 2009). Pet ownership decreases blood pressure responses to mental stress (Allen et al. 2001).

1.12 Light

Numerous designs for a Mars colony have proposed underground structures to avoid the cosmic radiation (Sheshpari et al. 2017). Underground domiciles would obviously negatively affect our exposure to natural sunlight. Divorcing or minimizing exposure to natural light and the kind and intensity of artificial light to replace the natural light will have behavioral effects that choice architects must contemplate. Alternate designs propose using ice or other transparent structures above ground that would shield from radiation but permit visible light to be let in (Sheshpari et al. 2017).

As far as choices involving interpersonal dynamics, dim or dark lighting increases feelings of anonymity, which may encourage risky or deviant behaviors (Haney et al. 1973).

On an individual level, low levels of light intensity can correlate with depressed moods, or even manifest into depressive disorders such as seasonal affective disorder (Begeman et al. 1997, Espiritu et al. 1994, Veitch 2011). These psychological states can obviously detrimentally affect proactive decision-making. Light intensity can be problematic with artificial light sources. One study showed that occupants of buildings exposed to daylight often received 2,000 or more lux (lumens per meter), whereas those in electrically lit buildings rarely received more than 100 lx (Cawthorne 1991). However, electrical light of sufficient intensity and wavelength can have biological and psychological effects comparable or similar to daylight (Lockley 2009). Light intensity is also important for maintaining the circadian rhythms that support sleep and a host of other biological functions (Cajochen et al. 2010).

Lack of sleep negatively affects mood, willpower, and cognitive tasks (Pilcher and Huffcutt 1996; Karatsoreos et al. 2011; Edlund 2011). Study participants who have missed just one night of sleep have poor impulse control indicative of a decrease in willpower: they take more risks in experimental gambling sessions and they eat more unhealthy food when given the opportunity compared to those who are not sleep-deprived (Venkatraman et al. 2011; Pardi et al. 2017; Greer et al. 2013). A single night of sleep deprivation increases the stress hormone cortisol by more than 100%. Cortisol increases appetite and preferentially leads to the storing of fat viscerally. In the evening after a night of sleep deprivation, cortisol still remains approximately 40% higher than normal (Basta et al. 2007).

Sleep loss also increases appetite by increasing levels of the hunger hormone ghrelin and decreasing levels of the satiety hormone leptin (Spiegel et al. 2004). In one study where participants were allowed to sleep only 4 h a night, ghrelin levels increased by 28% and leptin levels decreased by 18%. As a result, the study participants experienced a 24% increase in overall hunger, and a 32% increase in appetite for carbohydrates (Leproult and Van Cauter 2010).

For eons, our ancestors had eaten and engaged in other behaviors based on the light and dark cycles set by the sun. Our internal biological clock is located in the suprachiasmatic nucleus (SCN), which is situated in the hypothalamus of our brains. The SCN evolved in the time of our ancestors to respond to cues based on light or the lack of light from the sun, and prepare our bodies through changes in core body temperature, metabolism, and oscillations in gene expression and hormone production, for the beginnings of day or the coming of night. Those who could physiologically prepare ahead of time for day or night had an evolutionary advantage, and certain behaviors such as the timing of feeding evolved to be controlled by our SCN to optimally occur when we were awake during the hours in which our planet was illuminated by the light of the sun. The SCN keeps approximately thirty-seven trillion cells in our bodies in synchrony with one another (Saper et al. 2005, Bianconi et al. 2013). The pineal gland receives light input from the retina of the eye, and as it receives less light as night approaches, it secretes more and more of the sleep-inducing hormone melatonin (Brainard et al. 2001, McIntyre et al. 1989). Additionally, as the day goes on and you exert yourself, you break down the energy molecule adenosine triphosphate, or ATP, resulting in a buildup of the sleep-inducing byproduct adenosine. The sleep system and the arousal system are the yin and yang of sleep and wakefulness. They are each designed to inhibit the other, rather than co-exist (Saper et al. 2005). When lack of proper intensity and wavelength of light interferes with the production of melatonin, or lack of exercise results in less production of adenosine, sleep is disrupted (Chang et al. 2015).

For all these reasons, lighting on a Mars mission and colony should be designed to optimize sleep. Since sleep proceeds best in complete darkness, lights from electronic devices should be covered. During the evening hours, artificial light intake should be kept at a minimum to facilitate the production of melanin and other physiological changes to trigger your sleep. A study of the sleep patterns of 21 astronauts over 3,248 days of long-duration spaceflight on the space station, including 11 days prior to launch, which used physiological measuring devices and reviewed sleep logs to determine sleep medication use and sleep quality, revealed that sleep was often disrupted due to electronic tablet use before attempting to sleep (Flynn-Evans et al.

2016). A white noise producing source can mask distracting sounds. Sleep disruption can be minimized by limiting beverage intake in the hours before sleep, including stimulants (Haynes et al. 1975; Borkovec and Fowles 1973; Borkovec and Weerts 1976; Geer and Katkin 1966; Jacobson 1938; Kayumov et al. 2005).

Rational environmental design which facilitates these and other sleep-promoting behaviors can avoid the poor decision-making resulting from depressed mood, willpower, and cognitive function from deficient sleep.

1.13 Windows

Engineers involved in the early design of space capsules opposed providing any windows at all, because they presumed windows created unnecessary risks regarding atmospheric containment and only yielded when the first astronauts insisted on them (Wolfe 1979). Haines (1988) lauded the psychological value of windows in confined spaces in his paper on the design of space capsules (Haines 1988). In the interior of a confined room inside a space vehicle or colony habitat, the eye may never have an object on which to focus on that is more than a few meters away. At close focus, the ciliary muscles in the eye contract to thicken the lens and shorten focal length for viewing, which can cause tension and lead to headaches (Wolff et al. 2001). These can obviously lead to increased stress, and impairment of cognitive function and decision-making. Windows, which allow focus to expand toward distant objects, can relax the ciliary muscles and relieve stress.

Windows also let in direct and reflected sunlight which can assist all the biological functions dependent on circadian rhythms discussed in the previous section.

In the novel environment of space or a Mars colony, windows help establish psychological bearings by providing information about location, surroundings, and weather. Windows also can help alleviate feelings of isolation and boredom by providing additional stimuli to divert one's focus toward (Heerwagen 1990).

Windows can also enhance interpersonal relations and alleviate strains by promoting social distance and spaciousness which result in a decreased perception of crowding (Harrison et al. 1988; Desor 1972; Schiffenbauser et al. 1977).

1.14 Noise

Noise is often defined as unwanted sound (Lipscomb 1974; Schmidt, 2005). The intensity of sounds is typically measured on a decibel (db) scale. Measurements are usually made in dBA, a logarithmic scale. Long-term exposure to sounds of 85 dBA or more results in a loss of sensitivity to sounds (Stansfeld and Matheson 2003). Noise exposure has also been linked to nausea and headaches (Stansfeld and Matheson 2003; Evans 2001a).

Apart from these physical effects, noise can significantly affect an individual's psychology.

Prolonged exposure to noise increases stress and elevates blood pressure and stress hormones such as cortisol (Babisch et al. 2001; Babisch et al. 2003; Evans 2001b; Evans and Johnson 2000; Ising and Kruppa 2004; Stansfeld and Matheson 2003). The controllability and predictability of noise are more determinant factors in the creation of stress than even loudness (Glass and Singer 1972). Exposure to noise of 50 dBA or higher can increase heart rate and blood pressure during sleep, making sleep less restful (Stansfeld and Matheson 2003; Fritschi et al. 2011), adding to stress.

Noise also detrimentally interferes with the performance of tasks, especially when the tasks are demanding and complex (Smith 1989; Stansfeld and Matheson 2003).

On an interpersonal level, noise reduces the likelihood that someone will notice when others are in distress, and it also lessens the chances that they will choose to offer assistance if they do notice a problem (Jones et al. 1981; Mathews and Canon 1975).

For all these reasons, it is important either through wearable technology or sound-proofing of structures that noise levels are kept under control during the voyage to Mars and on a Mars colony.

1.15 Temperature

Highs and lows of temperature have been linked to aggression and increased stress (Anderson et al. 2000; Anderson and DeNeve 1992; Bell and Baron 1977; Haertzen et al. 1993). As would be expected, numerous studies show that multiple stressors can work synergistically to increase stress levels (Evans et al. 1996; Evans and Stecker 2004). Therefore, an omnipresent discomfort due to high or low temperature will add to all the other stressors on a Mars voyage and in a colony. To minimize this stress, internal temperature conditions should be optimized.

1.16 Odors and Contaminants

Efforts have been made on the International Space Station to remove malodorous smells which can increase aggravation and stress (Domanico 2014). Odors may emanate from equipment off-gassing, crew metabolic processes and excretions in clothes due to exercise, food, and experiments. Machinery such as a micropurification unit that is utilized on the International Space Station can remove both low and high molecular weight contaminants (Domanico 2014).

On a Mars mission and colony, in addition to simply removing odors to alleviate stress, odors can be proactively engineered within surroundings to produce beneficent psychological effects.

Inhaling green odor (a 50:50 mixture of trans-2-hexenal and cis-3-hexenol) has been shown to have an inhibitory effect on the stress-induced activation of the hypothalamo–pituitary–adrenocortical axis in humans (Oka et al. 2008), and to alleviate stress-induced cardiovascular, hormonal, and behavioral responses in rats (Ito et al. 2009; Nikaido and Nakashima 2009).

Furthermore, researchers found that rats who were experimentally induced to have depressive states through two behavioral models of depression had ameliorated depressive symptoms upon inhaling green odor, compared to rats that did not (Watanabe et al. 2011). Rats who inhaled the green odor had increased brain-derived neurotrophic factor (BDNF) in hippocampal regions. Increased neural connections due to stimulation by BDNF are thought to be necessary for antidepressants which increase monoamine levels to exert their effects.

Other fragrances have also been reported to alleviate symptoms of stress and exert calming effects such as coconut (Mezzacappa et al. 2010) and rose odors (Fukada et al. 2007).

Numerous contaminants, termed endocrine disrupters, can mimic or disrupt hormones and thus affect behavioral choices. In addition to machinery to remove particles that cause aggravating odors, NASA scientists have found that certain plants can remove toxins from the air, such as gerbera daisy and chrysanthemum (to remove benzene), elephant ear and golden pothos (to remove carbon monoxide), and heart leaf and aloe vera (to remove formaldehyde) (Dingle et al. 2000). Perhaps, other plants may prove useful to remove various endocrine disrupters.

1.17 Order

The broken windows theory is an environmental psychology theory that postulates that people will act more unruly in disordered environments. Broken window theory experiments by psychologist Philip Zimbardo include one where two cars were left in the same city; one clean and in working order, and the other in disarray (Wilson and Kelling 1982). The ordered car was left alone for more than a week but the disordered car was stripped by vandals in a matter of a few hours. It was only when the ordered car's window was broken by the experimenters that it too was vandalized in a matter of hours.

The theory, like other theories of environmental psychology, postulates that people look to their surrounding environment for signals about how to behave. A disordered environment conveys the message that a social norm exists where disruptive behavior is tolerated.

To avoid this message on a Mars mission and colony, norms should be established before the mission about the disposal of liter, the drying of laundry, and the general order of items in rooms, especially public spaces. On the International Space Station, astronaut Clayton Anderson remarked that it was up to each astronaut where to store their sweaty workout gear after exercise and some astronauts stored gear in places where odor spread to the rest of the crew (Domanico 2014).

1.18 Diversions

Psychological diversions on the Mars mission and colony are a must to avoid the boredom concomitant with voyages of significant duration and an atmosphere that prohibits the freedom of movement enjoyed on Earth.

Boredom has been divergently conceptualized in the research literature as a stressor which causes increased cortisol and heart rate (Merrifield and Danckert 2013) and as a depressor of physiology that does not result in hypothalamic–pituitary–adrenal activation integral to the stress response (Thackray 1981). Regardless of this divergence, it is generally accepted that repetitive tasks and a lack of novel stimulation cause boredom, and this boredom can result in negative affect (James 1982; Rupp and Vodanovich 1997).

In the arrangement of diversions in public spaces on a Mars mission in the form of games and other activities, those that involve human touch will likely be important for the satisfaction of relationship needs. Some nonhuman primates spend up to 20% of their time grooming to facilitate interpersonal bonds, and in a recent study of sports teams, those that spend more time touching one another perform more successfully (Kraus et al. 2010). Touch dampens the stress response and increases levels of oxytocin while decreasing blood pressure (Coan et al. 2006; Light et al. 2004). Therefore, activities involving touches such as partner yoga, dance, martial arts, or massage may be beneficial on a Mars mission. Perhaps, some of these activities, which may include games such as a three dimensional kind of twister, can be enhanced through the use of augmented reality.

1.19 Conclusion

No matter how far we soar into the stars, our psychologies will be ever bound to the totality of our surroundings. The uncanny Martian surroundings and great expanse of space to voyage there will be mismatched with the environment in which our genetics, epigenetics, and psyches evolved to a degree never before encountered by *Homo sapiens*. Within this mismatch, the first voyagers to Mars will encounter a space mission of unprecedented duration, and alienation from our terrestrial home. On this strange journey and land, psychological vulnerabilities will be magnified and the satisfaction of fundamental human existential, relationship, security, and fitness needs will be challenged. To decrease the obstacles to the satisfaction of these challenges, it would be prudent to engage in pre-mission planning to engineer living environments that optimize human factors. Aspects of our environmental surroundings that should receive pre-mission attention include primes, defaults, private spaces, shared spaces, ceiling height, object shape, color, nature, pets, light, windows, noise, temperature, odors, contaminants, order, and diversions. The breadth of these environmental challenges that can affect our psychologies reveal what a fragile species we are. But we have the spirit of voyagers. This exploratory imperative and the will to transcend the comforts of our environment necessitate that we adapt, the great strength of our kind. To adapt to the first journey where we seek to live beyond the soil of our terrestrial home, we must shape our psyches by shaping our environment. Only then can we shape our future.

References

Alexander, C., Ishikawa, S., Silverstein, M., Jacobson, M., Fiksdahl-King, I., & Angel, S. (1977). *A pattern language: Towns, buildings, construction*. Oxford University Press.

Allen, K., Shykoff, B., & Izzo, J. (2001). Pet ownership, but not ace inhibitor therapy, blunts home blood pressure responses to mental stress. *Hypertension, 38*(4), 815–820.

Althoff, T., White, R., & Horvitz, E. (2016). Influence of Pokemon go on physical activity: Study and implications. *Journal of Medical Internet Research, 18*(1), e315.

Altman, I. (1981). *Environment and social behavior: Privacy, personal space, territory, and crowding*. New York: Irvington Press.

Altman, I., & Haythorn, W. (1967). The ecology of isolated groups. *Behavioral Sciences & The Law, 12*(3), 169–182.

Anderson, C., & DeNeve, K. (1992). Temperature, aggression, and the negative affect escape model. *Psychological Bulletin, 111*(2), 347–351.

Anderson, C., Anderson, K., Dorr, N., DeNeve, K., & Flanagan, M. (2000). Temperature and aggression. In M. P. Zanna (Ed.), *Advances in experimental social psychology* (Vol. 32, pp. 63–133). San Diego, CA: Academic Press.

Ardrey, R. (1967). *The territorial imperative: A personal inquiry into the animal origins of property and nations*. London: Collins.

Arzy, S., Seeck, M., Ortigue, S., Spinelli, L., & Blanke, O. (2006). Induction of an illusory shadow person. *Nature, 443,* 287.

Babisch, W., Fromme, H., Beyer, A., & Ising, H. (2001). Increased catecholamines levels in urine in subjects exposed to road traffic noise: The role of stress hormones in noise research. *Environment International, 26*(7–8), 475–481.

Babisch, W., Segal, S., Eviatar, E., Lapinsky, J., Shlamkovitch, N., & Kessler, A. (2003). Stress hormones in the research on cardiovascular effects of noise. *Stress, 5*(18), 1–11.

Bachman, K., Otto, C., & Leveton, L. (2012). Countermeasures to mitigate the negative impact of sensory deprivation and social isolation in long-duration space flight. In NASA/TM-2012-217365.

Bargh, J., Chen, M., & Burrows, L. (1996). Automaticity of social behavior: Direct effects of trait construct and stereotype-activation on action. *Journal of Personality and Social Psychology, 71*(2), 230–244.

Bargh, J., Gollwitzer, P., Lee-Chai, A., Barndollar, K., & Troestschel, R. (2001). The automated will: Nonconscious activation and pursuit of behavioral goals. *Journal of Personality and Social Psychology, 81,* 1014–1027.

Barnes, K., & Gibson, N. (2013). Supernatural agency: Individual difference predictors and situational correlates. *International Journal for the Psychology of Religion, 23*(1), 42–62.

Basta, M., Chrousos, G., Vela-Bueno, A., & Vgontzas, A. (2007). Chronic insomina and stress system. *Sleep Medicine Clinics, 2*(2), 279–291.

Begeman, S., Van den Beld, G., & Tenner, A. (1997). Daylight, artificial light and people in an office environment, overview of visual and biological responses. *International Journal of Industrial Ergonomics, 20*(3), 231–239.

Bell, P., & Baron, R. (1977). Aggression and ambient temperature: The facilitating and inhibiting effects of hot and cold environments. *Bulletin of the Psychonomic Society, 9*(6), 443–445.

Bentham, J. (1798). *Proposal for a new and less expensive mode of employing and reforming convicts*. London.

Benton, M. (2003). *When life nearly died: The greatest mass extinction of all time*. New York: Thames & Hudson.
Bexton, W., Heron, W., & Scott, T. (1954). Effects of decreased variation in the sensory environment. *Canadian Journal of Psychology, 8,* 70–76.
Bianconi, E., Piovesasn, A., Facchin, F., Beradui, A., Casedei, R., Frabetti, F., et al. (2013). An estimation of the number of cells in the human body. *Annals of Human Biology, 40*(6), 463–471.
Borkovec, T., & Fowles, D. (1973). Controlled investigation of effects of progressive and hypnotic relaxation on insomnia. *Journal of Abnormal Psychology, 82,* 153.
Borkovec, T., & Weerts, T. (1976). Effects of progressive relaxation on sleep disturbance: An electroencephalographic evaluation. *Psychosomatic Medicine, 33,* 173.
Brainard, G., Hanifin, J., Greeson, J., Byrne, B., Glickman, G., Gerner, E., et al. (2001). Action spectrum for Melatonin regulation in humans: Evidence for a novel circadian photoreceptor. *Journal of Neuroscience, 21*(16), 6405–6412.
Braum, A., & Paulus, P. (1987). Crowding. In D. Stokols & I. Altman (Eds.), *Handbook of environmental psychology* (pp. 533–570). New York: Wiley.
Bridges-Parlet, S., Knopman, D., & Thompson, T. (1994). A descriptive study of physically aggressive behavior in dementia by direct observation. *Journal of American Geriatric Society, 42*(2), 192–197.
Broady, M. (1966). Social theory in architectural design. *Arena, 81,* 149–154.
Brown, B. (1987). Territoriality. In D. Sotkols & I. Altman (Eds.), *Handbook of environmental psychology* (pp. 505–531). New York: Wiley.
Budde, H., Voelcker-Rehage, C., Pietrabyk-Kendziorra, S., Ribeiro, P., & Tidow, G. (2008). Acute coordinative exercise improves attentional performance in adolescents. *Neuroscience Letters, 441*(2), 219–223.
Burke, P. (1996). Social identities and psychosocial stress. In H. B. Kaplan (Ed.), *Psychosocial stress: Perspectives on structure, theory, life-course, and methods* (pp. 141–174).
Cajochen, C., Chellappa, S., & Schmidt, C. (2010). What keeps us awake?–The role of clocks and hourglasses, light, and melatonin. *International Review of Neurobiology, 93,* 57–90.
Cawthorne, D. (1991). *Buildings lighting and the biological clock, martin centre for architectural and urban studies*. Cambridge: University of Cambridge.
Chang, C., Lin, Y., & Chou, M. (2008). Experiences and stress reduction of viewing natural environmental settings. *Acta Horticulturae, 775,* 139–146.
Chang, A., Aeschbach, D., Duffy, J., & Czeisler, C. (2015). Evening use of light-emitting eReaders negatively affects sleep, circadian timing, and next-morning alertness. *Proceedings of the National Academy of the Sciences of the United States of America, 112*(4), 1232–1237.
Charnetski, C., Riggers, S., & Brennan, F. (2004). Effect of petting a dog on immune system function. *Psychological Reports, 95*(3), 1087–1091.
Coan, J. A., Schaefer, H. S., & Davidson, R. J. (2006). Lending a hand: Social regulation of the neural response to threat. *Psychological Science, 17*(12), 1032–1039.
Cockell, C. (2002). Mars is an awful place to live. *Interdisciplinary Science Reviews, 27*(1), 32–38.
Cole, S., Capitanio, J., Chun, K., Arevalo, J., Ma, J., & Cacioppo, J. (2015). Myeloid differentiation architecture of leukocyte transcriptome dynamics in perceived social isolation. *PNAS, 112*(49), 15142–15147.
Cooley, D., Jewitt, B., & Oren, E. (1973). Territoriality–Its effects on the correctional environment (NCJ Number: 28207). Canadian Solicitor General.
Curran, S., Blatchley, R. J., & Hanlon, T. E. (1978). The relationship between body buffer zone and violence as assessed by subjective and objective techniques. *Criminal Justice & Behavior, 5*(1), 53–62.
D'Atri, D., & Ostfield, A. (1975). Crowding: Its effects on the elevation of blood pressure in a prison setting. *Preventative Medicine, 4,* 550–566.
Daffern, M., Mayer, M., & Martin, T. (2004). Environment contributors to aggression in two forensic psychiatric hospitals. *International Journal of Forensic Mental Health, 3*(1), 105–114.

Daniel, M., Martin, A., & Carter, J. (1992). Opiate receptor blockade by naltrexone and mood state after acute physical activity. *British Journal of Sports Medicine, 26,* 111–115.
D'Ausilio, A., Novembre, G., Fadiga, L., & Keller, P. (2015). What can music tell us about social interaction? *Trends in Cognitive Science, 19*(3), 111–114.
Dazkir, S., & Read, M. (2011). Furniture forms and their influence on our emotional responses toward interior environments. *Environment and Behavior, 44*(5), 722–732.
Desor, J. (1972). Toward a psychological theory of crowding. *Journal of Personality and Social Psychology, 21*(1), 79083.
Diette, G., Lechtzin, N., Haponik, E., Devrotes, A., & Rubin, H. (2003). Distraction therapy with nature sights and sounds reduces pain during flexible bronchoscopy: A complementary approach to routine analgesia. *Chest, 123*(3), 941–948.
Dingle, P., Tapsell, P., & Hu, S. (2000). Reducing formaldehyde exposure in office environments using plants. *Bulletin of Environmental Contamination and Toxicology, 64*(2), 302–308.
Domanico, A. (2014, October 22). This is how NASA banishes body odors in space. *Cnet.*
Dutton, D. (2003). Aesthetics and evolutionary psychology. In J. Levinson (Ed.), *The oxford handbook for aesthetics.* Oxford University Press.
Edlund, M. (2011, April 7). Is depression making me sleepless, or is insomnia making me depressed? *Psychology Today.*
Ekman, P., & Friesen, W. (1975). *Pictures of facial affect.* Palo Alto: Consulting Psychologists Press.
Elliot, A., & Maier, M. (2003). Color psychology: Effects of perceiving color on psychological functioning in humans. *Annual Review of Psychology, 65*(1), 95–120.
Erisman, J., Sutton, M., Galloway, J., Klimont, Z., & Winiwarter, W. (2008). How a century of ammonia synthesis changed the world. *Nature Geoscience, 1,* 636.
Espiritu, R., Kripke, D., Ancoli-Israel, S., Mowen, M., Mason, W., & Fell, R. (1994). Low illumination experienced by San Diego adults: Association with atypical depressive symptoms. *Biological Psychiatry, 35*(6), 403–407.
Evans, G. (Ed.). (1982). *Environmental stress.* New York: Cambridge University Press.
Evans, G. (2001a). Environmental stress and health. In A. Baum & T. Revenson (Eds.), *Handbook of health psychology* (pp. 365–385). Hillsdale, NJ: Erlbaum.
Evans, G. (2001b). Environment stress and health. In A. Baum & T. Revenson (Eds.), *Handbook of health psychology* (pp. 365–385). Hillsdale, NJ: Erlbaum.
Evans, G., & Johnson, D. (2000). Stress and open-office noise. *Journal of Applied Psychology, 85*(5), 779–783.
Evans, G., & Stecker, R. (2004). The motivational consequences of environmental stress. *Journal of Environmental Psychology, 24,* 143–165.
Evans, G., Allen, K., Tafalla, R., & O'Meara, T. (1996). Multiple stressors. *Journal of Environmental Psychology, 16,* 147–154.
Fagan-Pryor, E., Haber, L., Dunlap, D., Nall, J., Stanley, G., & Wolpert, R. (2003). Patients' views of causes of aggression by patients and effective interventions. *Psychiatric Services, 54,* 549–553.
Festinger, L., Schachter, S., & Back, K. (1950). *Social pressures in informal groups: A study of human factors in housing.* Stanford, CA: Stanford University Press.
Flynn-Evans, E., Barger, L., Kubey, A., Sullivan, J., & Czeisler, C. (2016). Circadian misalignment affects sleep and medication use before and during spaceflight. *NPJ Microgravity,* 15019.
Foucault, M. (1979). *Discipline and punishment: The birth of the prison.* New York: Vintage Press.
Friedmann, E., Katcher, A., Lynch, J., & Thomas, S. (1980). Animal companions and one-year survival of patients after discharge from a coronary care unit. *Public Health Reports, 95*(4), 307–312.
Fritschi, L., Brown, A., Kim, R., Schwela, D., & Kephalopoulos, S. (Eds.). (2011). *Burden of disease from environment from environmental noise: Quantification of healthy life years lost in Europe.* Bonn, Germany: WHO European Centre for Environment and Health.
Frone, M., Russell, M., & Cooper, M. (1991). Relationship of work and family stressors to psychological distress: The independent moderating influence of social support, mastery, active coping, and self-focused attention. *Journal of Social Behavior & Personality, 6*(7), 227–250.

Fukada, M., Kaidoh, T., Ito, A., Yano, T., Hayashibara, C., & Watanabe, T. (2007). "Green odor" inhalation reduces the skin-barrier disruption induced by chronic restraint stress in rats: Physiological and histological examinations. *Chemical Senses,* 32633–32639.

Geer, J., & Katkin, E. (1966). Treatment of insomnia using a variant of systematic desensitization. *Journal of Abnormal Psychology, 71,* 161.

Glass, D., & Singer, J. (1972). *Urban stress.* New York: Academic Press.

Grassian, S., & Friedman, N. (1986). Effects of sensory deprivation in psychiatric seclusion and solitary and solitary confinement. *International Journal Law & Psychiatry, 8*(1), 49–65.

Greer, S., Goldstein, A., & Walker, M. (2013). The impact of sleep deprivation on food desire in the human brain. *Nature Communications, 4,* 2259.

Haertzen, C., Buxton, K., Covi, L., & Richards, H. (1993). Seasonal changes in rule infractions among prisoners: A preliminary test of the temperature-aggression hypothesis. *Psychological Reports, 72*(1), 195–2000.

Haines, R. (1988). Space station proximity operations and window design. *Space Station Human Factors Research Review, 4,* 1–18.

Haney, C. (2003). Mental health issues in long-term solitary and "supermax" confinement. *Crime and Delinquency, 49*(1), 124–156.

Haney, C., Banks, C., & Zimbardo, P. (1973). Interpersonal dynamics in a simulated prison. *International Journal of Criminology and Penology, 1,* 69–97.

Harrison, A., Caldweli, B., Struthers, N., & Clearwater, Y. (1988). *Incorporation of privacy elements in space station design.* Washington, DC: NASA.

Harrison, A., Clearwater, Y., McKay, C. (Eds.). (1991). *From Antarctica to outer space: Life in isolation and confinement.* Springer.

Hartig, T., Evans, G., Jamner, L., David, D., & Garling, T. (2003). Tracking restoration in natural and urban field settings. *Journal of Environmental Psychology, 23*(2), 109–123.

Haynes, S., Price, M., & Simons, J. (1975). Stimulus control treatment of insomnia. *Journal of Behavior Therapy and Experimental Psychiatry, 6,* 279.

Heerwagen, J. (1990). The psychological aspects of windows and window design. In R. Selby & K. Anthony (Eds.), *Coming of age* (pp. 269–280). Oklahoma City, OK: Environmental Design Research Association.

Hildreth, A., Derogatis, L. R., & Mccusker, K. (1971). Body buffer zone and violence: A reassessment and confirmation. *American Journal of Psychiatry, 127*(12), 1641–1645.

Ising, H., & Kruppa, B. (2004). Health effects caused by noise: Evidence in the literature from the past 25 years. *Noise and Health, 6*(22), 5–13(19).

Ito, A., Miyoshi, M., Ueki, S., Fukada, M., Komaki, R., & Watanabe, T. (2009). "Green odor" inhalation by rats down-regulates stress-induced increases in Fos expression in stress-related forebrain regions. *Neuroscience Research, 65,* 166–174.

Ittelson, W., Proshansky, H., Rivlin, L., & Winkel, G. (1974). *An introduction to environmental psychology.* New York: Holt, Rinehart and Winston.

Jacobson, E. (1938). *Progressive relaxation.* Oxford, England: University Chicago Press.

James, A. (1982). Boredom–The stress of predictability. *Journal of Physical Education, Recreation & Dance, 53*(7), 35–36.

Johnston, N. (2000). *Forms of constraint: A history of prison architecture.* Urbana, Chicago: University of Illinois Press.

Jones, D., Chapman, A., & Auburn, T. (1981). Noise in the environment: A social perspective. *Journal of Environmental Psychology, 1*(1), 43–59.

Kahneman, D., Knetsch, J., & Thaler, R. (1990). Experimental tests of the endowment effect and the Coase Theorem. *Journal of Political Economy, 98,* 1325–1348.

Kansas, N. (2010). Expedition to mars: Psychological, interpersonal, and psychiatric issues. *Journal of Cosmology, 12,* 3741–3747.

Karatsoreos, I., Bhagat, S., Bloss, E., Morrison, J., & McEwen, B. (2011). Disruption of Circadian clocks has ramifications for metabolism, brain and behavior. *Proceedings of the National Academy of Sciences of the United States of America, 108*(4), 1657–1662.

Kayumov, L., Casper, R., Hawa, R., Perelman, B., Chung, S., Sokalsky, S., et al. (2005). Blocking low-wavelength light presents nocturnal melatonin suppression with no adverse effects on performance during stimulated shift work. *Journal of Clinical Endocrinology and Metabolism, 90*(5), 2755–2761.

Kinzel, A. (1970). Body-buffer zone in violent prisoners. *American Journal of Psychiatry, 127*(1), 59–64.

Knapp, M., Hall, J., & Horgan, T. (2014). The effects of the environment on human communication. In *Nonverbal communication in human interaction* (8th ed., pp. 89–122). Boston, MA: Wadsworth.

Koch, C. (2009). *Minds, brains, and society*. Pasadena, CA: Lecture at Caltech.

Koopmann, A., Leménager, T., Wolf, N., Reinhard, I., Hermann, D., Koch, J., et al. (2014). The impact of atrial natriuretic peptide on anxiety, stress and craving in patients with alcohol dependence. *Alcohol and Alcoholism, 49*(3), 282–286.

Korpela, K. (1992). Adolescents' favourite places and environmental self-regulation. *Journal of Environmental Psychology, 12*(3), 249–258.

Korpela, K., & Hartig, T. (1996). Restorative qualities of favorite place. *Journal of Environmental Psychology, 16*(3), 221–233.

Korpela, K., Hartig, T., Kaiser, F., & Fuhrer, U. (2001). Restorative experience and self-regulation in favorite places. *Environment and Behavior, 33*(4), 572.

Kraus, M., Huang, C., & Keltner, D. (2010). Tactile communication, cooperation, and performance: An ethological study of the NBA. *Emotion, 10,* 745–749.

Kraut, R., Edigo, C., & Galegher, J. (1990). Patterns of contact and communication in scientific research collaboration. In *Intellectual teamwork: Social and technological foundations of cooperative work* (pp. 149–171). Hillsdale, NJ: Lawrence Erlbaum Associates.

Kuo, F., & Sullivan, W. (2001). Aggression and violence in the inner city: Effects of environment via mental fatigue. *Environment and Behavior, 33*(4), 543.

Kurosawa, A. (1954). *The seven samurai* script, reprinted in Kurosawa, A., Richie, D. (1992). *The seven samurai: And other screenplays*. New York: Faber & Faber.

Kweon, B., Ulrich, R., Walker, V., & Tassinary, L. (2008). Anger and stress: The role of landscape posters in an office setting. *Environment and Behavior, 40,* 355–381.

Langer, E., & Saegert, S. (1977). Crowding and cognitive control. *Journal of Personality and Social Psychology, 35*(3), 175–182.

Lawrence, C., & Andrews, K. (2004). The influence of perceived prison crowding on male inmates perception of aggressive events. *Aggressive Behavior, 30*(4), 273–283.

Leather, P., Pyrgas, M., Beale, D., & Lawrence, C. (1998). Windows in the workplace: Sunlight, view and occupational stress. *Environment and Behavior, 30*(6), 739.

Leproult, R., & Van Cauter, E. (2010). Role of sleep and sleep loss in hormonal release and metabolism. In S. Loche, M. Cappa, L. Ghizzoni, M. Maghnie, M. Savage (Eds.), *Pediatric neuroendocrinology: Endocrine development* (Vol. 17, pp. 11–21).

Li, Q., Morimoto, K., Kobayashi, M., Inagaki, H., Katsumata, M., Hirata, Y., et al. (2008). Visiting a forest, but not a city, increases human natural killer activity and expression of anti-cancer proteins. *International Journal of Immunopathology and Pharmacology, 21*(1), 117–127.

Liederman, P. (1962). Man alone: Sensory deprivation and behavior change. *Correctional Psychiatry and Journal of Social Therapy, 8,* 63–74.

Light, K., Grewen, K., Amico, J. (2004). More frequent partner hugs and higher oxytocin levels are linked to lower blood pressure and heart rate in premenopausal women. *Biological Psychology, 69,* 5–21.

Lilly, J. (1956). Mental effects of reduction of ordinary levels of physical stimuli on intact healthy persons. *Psychological Research Reports, 5,* 1–9.

Lin, T., & Kuo, Y. (2013). Exercise benefits brain function: The monoamine connection. *Brain Science, 3*(1), 39–53.

Lipscomb, D. (1974). Noise: The unwanted sounds. Chicago: Nelson-Hall.

Lockley, S. (2009). Circadian rhythms: Influence of light in humans. *Encyclopedia of Neuroscience*, 971–988.
Lohr, V. (2007). Benefits of nature: What we are learning about why people respond to nature. *Journal of Physiological Anthropology, 26*(2), 83–85.
Losey, R., Bazaliiskii, V., Garvie-Lok, S., Germonpré, M., Leonard, J., Allen, A., et al. (2011). Canids as persons: Early Neolithic dog and wolf burials, Cis-Baikal, Siberia. *Journal of Anthropological Archaeology, 30*(2), 174–189.
Maier, S., & Seligman, M. (1976). Learned helplessness: Theory and evidence. *Journal of Experimental Psychology, 105*, 3–46.
Maslow, A. (1943). A theory of human motivation. *Psychological Review, 50*(4), 370–396.
Mathews, K., & Canon, L. (1975). Environmental noise level as a determinant helping behavior. *Journal of Personality and Social Psychology, 32*(4), 571–577.
McIntyre, I., Norman, T., Burrows, G., & Armstrong, S. (1989). Humans melatonin suppression by light is intensity dependent. *Journal of Pineal Research, 6*(2), 149–156.
Mehta, R., & Zhu, R. (2009). Blue or red? Exploring the effect of color on cognitive task performances. *Science, 323*(5918), 1226–1229.
Mequida, C., & Wiener, N. (1996). Human collective aggression: A behavioral ecology perspective. *Ethology and Sociobiology, 17*(4), 247–262.
Merrifield, C., & Danckert, J. (2013). Characterizing the psychophysiological signature of boredom. *Experimental Brain Research, 232*(2).
Meyers-Levy, J., & Rui, Z. (2007). The influence of ceiling height: The effect of priming on the type of processing that people use. *Journal of Consumer Research, 34*(2), 174–186.
Mezzacappa, E., Arumugam, U., Chen, S., Stein, T., Oz, M., Buckle, J. (2010). Coconut fragrance and cardiovascular response to laboratory stress: Results of pilot testing. *Holistic Nursing Practice*, 24322–24332.
Milgram, S. (1970). The experience of living in cities. *Science, 167*, 146–1468.
Mooney, P., & Nicell, P. (1992). The importance of exterior environment for Alzheimer residents: Effective care and risk management. *Health Manage Forum, 5*(2), 23–29.
Moore, E. (1985). *Environmental variables affecting prisoner health care demands*. Paper presented at the Proceedings of the American Institute of Architect, Los Angeles.
Moscicka, A., Czarnecka, K., & Ciach, T. (2007). IntelliDrug implant for medicine delivery in Alzheimer's disease treatment. *Macromolecular Symposia, 253*(1), 134–138.
Murphy, M., Slavich, G., Rohleder, N., & Miller, G. (2012). Targeted rejection triggers differential pro- and anti-inflammatory gene expression in adolescents as a function of social status. *Clinical Psychological Science, 1*(1), 30–40.
Nagasawa, M., Kikusui, T., Onaka, T., & Ohta, M. (2009). Dog's gaze at its owner increases owner's urinary oxytocin during social interaction. *Hormones and Behavior, 55*(3), 434–441.
Nanden, U., Eisen, S., & Baladandayuthapani, V. (2008). Undertaking an art survey to compare patient versus student art preferences. *Environment and Behavior, 40*(2), 269.
National Institute of Mental Health. https://www.nimh.nih.gov/health/statistics/suicide.shtml.
Nelson, M., Gray, K., & Allen, J. P. (2015). Group dynamics challenges: Insights from biosphere 2 experiments. *Life Sciences in Space Research, 6*, 79–86.
Nijman, H., Campo, J., Ravell, D., & Merckelbach, H. (1999). A tentative model of aggression on inpatient psychiatric wards. *Psychiatric Services, 50*(6), 832–834.
Nikaido, Y., & Nakashima, T. (2009). Effects of environmental novelty on fear-related behavior and stress responses of rats to emotionally relevant odors. *Behavioural Brain Research, 199*, 241–246.
O'Neal, E., Brunault, M., Marquis, J., & Carifio, M. (1979). Anger and the body-buffer zone. *Journal of Social Psychology, 108*, 135–136.
Oka, T., Hayashida, S., Kaneda, Y., Takenaga, M., Tamagawa, Y., & Tsuji, S. (2008). Green odor attenuates a cold pressor test-induced cardiovascular response in healthy adults. *Biopsychosocial Medicine, 2*, 210.

Paluck, R. J., & Esser, A. (1971). Controlled experimental modification of aggressive behavior in territories of severely retarded boys. *American Journal of Mental Deficiency, 76*(1), 23–29.
Pardi, D., Buman, M., Black, J., Lammers, G., & Zeitzer, J. (2017). Eating decisions based on alertness levels after a single night of sleep manipulation: A randomized clinical trial. *Sleep, 40*(2).
Park, S., & Mattson, R. (2008). Effects of flowering and foliage plants in hospital rooms on patients recovering from abdominal surgery. *HortTechnology, 18,* 4563–4568.
Paulus, P. B. (1988). *Prison crowding: A psychological perspective (research in criminology).* New York: Springer.
Pegna, A., et al. (2005). Discriminating emotional faces without primary visual cortices involves the right amygdala. *Nature Neuroscience, 8*(1), 24–25.
Pilcher, J., & Huffcutt, A. (1996). Effects of sleep deprivation on performance: A meta-analysis. *Sleep, 19*(4), 318–326.
Poli, E., & Angrilli, A. (2015). Greater general startle reflex is associated with greater anxiety levels: A correlational study on 111 young women. *Frontiers in Behavioral Neuroscience, 9,* 10.
Pollack, M. (2012, June 15). Answering a question about an airport bathroom. *The New York Times.*
Ray, D., Wandersman, A., Ellisor, J., & Huntington, D. (1982). The effects of high density in a juvenile correctional institution. *Basic and Applied Social Psychology, 3,* 95–108.
Roger, D., & Schalekamp, E. (1976). Body-buffer zone and violence: A cross-cultural study. *Journal of Social Psychology, 98,* 153–158.
Roush, D. (1989). *Far from the maddening crowd: The Relationship between crowding and safety in Juvenile institutions.* Paper presented at the Congress of Corrections of the American Correctional Association, Baltimore, MD.
Rupp, D., & Vodanovich, S. (1997). The role of boredom proneness in self-reported anger and aggression. *Journal of Social Behavior & Personality, 12*(4), 925–936.
Ryden, M., Bossenmaier, M., & McLachlan, C. (1991). Aggressive behavior in cognitively impaired nursing home residents. *Research in Nursing & Health, 14*(2), 87–95.
Salgado, G. (2016, April 9). When the oil fields burned. *The New York Times.*
Samuelson, W., & Zeckhauser, W. (1988). Status quo bias in decision making. *Journal of Risk and Uncertainty, 1,* 7–59.
Saper, C., Scammell, T., & Lu, J. (2005). Hypothalamic regulation of sleep and circadian rhythms. *Nature, 437*(7063), 1257–1263.
Scheibehenne, B., Greifeneder, R., & Todd, P. (2010). Can there ever be too many options? A meta-analytic review of choice overload. *Journal of Consumer Research, 37*(3), 409–425.
Schiffenbauser, A., Brown, J., Perry, P., Shulack, L., & Zanzola, A. (1977). The relationship between density and crowding: Some architectural modifiers. *Environment and Behavior, 9*(1), 3.
Schmidt, C. (2005). Noise that annoys: Regulating unwanted sound. *Environmental Health Perspectives, 113*(1), A42.
Segal, M. (1974). Alphabet and attraction: An unobtrusive measure of the effect of propinquity in a field setting. *Journal of Personality and Social Psychology, 30,* 654–657.
Seligman, M. (2006). *Learned optimism.* Vintage.
Semple, J. (1993). *Bentham's prison: A study of the panopticon penitentiary*. New York, USA: Oxford University Press.
Sherman, S., Varni, J., Ulrich, R., & Malcarne, V. (2005). Post-occupancy evaluation of healing gardens in a pediatric cancer center. *Landscape and Urban Planning, 73*(2–3), 167–183.
Sheshpari, M., Fujii, Y., & Tani, T. (2017). Underground structures in Mars excavated by tunneling methods for sheltering humans. *Tunnelling and Underground Space Technology, 64,* 61–73.
Shimbun, Y. (2008, December 10). Blue streetlights believed to prevent suicides, street crime. *The Seattle Times.*
Sleiman, S., Henry, J., Al-Haddad, R., El Hayek, L., Abou Haidar, E., Stringer, T., et al. (2016). Exercise promotes the expression of brain derived neurotrophic factor (BDNF) through the action of the ketone body β-hydroxybutyrate. *Elife., 5,* e15092.
Smil, V. (1999). Detonator of the population explosion. *Nature, 400,* 415.

Smith, A. (1989). A review of the effects of noise on human performance. *Scandinavian Journal of Psychology, 30*(3), 185.

Spiegel, K., Tasali, E., Penev, P., & Van Cauter, E. (2004). Brief communication: Sleep curtailment in healthy young men is associated with decreased leptin levels, elevated ghrelin levels, and increased hunger and appetite. *Annals of Internal Medicine, 141,* 846–850.

Stansfeld, S., & Matheson, M. (2003). Noise pollution: Non-auditory effects on health. *British Medical Bulletin, 68*(1), 243–257.

Steptoe, A., Shankar, A., Demakakos, P., & Wardle, J. (2013). Social isolation, loneliness, and all-cause mortality in older men and women. *PNAS, 110*(15), 5797–5801.

Sundstrom, E., & Altman, I. (1972). *Relationships between dominance and territorial behavior: Field study in a young rehabilitation setting*. Washington, DC: Law Enforcement Assistance Administration. Technical Report (Grant No. 70-065-PG-21).

Sweeney, P., Anderson, K., & Bailey, S. (1986). attributional style in depression: A meta-analytic review. *Journal of Personality and Social Psychology, 50,* 974–991.

Szocik, K., Marques, E., Abood, S., Kedzior, A., Lysenko-Ryba, K., & Minich, D. (2018). Biological and social challenges of human reproduction in a long-term Mars base. *Futures, 100,* 56–62.

Thackray, R. (1981). The stress of boredom and monotony: A consideration of the evidence. *Psychosomatic Medicine, 43*(2).

Toro, R., et al. (2008). Brain size and folding of the human cerebral cortex. *Cerebral Cortex, 18*(10), 2352–2357.

Tversky, A., & Kahneman, D. (1974). Judgement under uncertainty: Heuristics and biases. *Science, 185,* 1124–1131.

Ulrich, R. (1979). Visual landscapes and psychological well-being. *Landscape research, 4*(1), 17–23.

Ulrich, R. (1984). View through a window may influence recovery from surgery. *Science, 224*(4647), 420–421.

Ulrich, R. (1999). Effects of gardens on health outcomes: Theory and research. In *Healing gardens: Therapeutic benefits and design recommendations*, 27–86.

Van den Berghe, P. L. (1974). Bringing beasts back in: Towards a biosocial theory of aggression. *American Sociological Review, 39*(6), 777–788.

Veitch, J. (2011). Workplace design contributions to mental health and well-being. *Healthcare Papers, 11,* 38–46.

Venkatraman, V., Huettel, S., Chuah, L., Payne, J., & Chee, M. (2011). Sleep deprivation biases the neural mechanisms underlying economic preferences. *Journal of Neuroscience, 31*(10), 3712–3718.

Vodanovich, S. (2003). Psychometric measures of boredom: A review of the literature. *Journal of Psychology: Interdisciplinary & Biological Processes, 66*(2), 146–167.

Wald, M. (1998, July 1). U.S. nuclear arms' cost put at $5.48 trillion. *The New York Times.*

Wallace, M. (2018, January 24) Tamagotchi returns again with cool retro colors and limited flocked variants. *Forbes.*

Watanabe, T., Fujihara, M., Murakami, E., Miyoshi, M., Tanaka, Y., Koba, S., et al. (2011). Green odor and depressive-like state in rats: Toward an evidence-based alternative medicine? *Behavioural Brain Research, 224*(2), 290–296.

Werner, R. (2012). *The environmental psychology of prisons and jails: Creating humane spaces in secure settings*. New York: Cambridge University Press.

Whitehead, D. (2008). The history of the Sims. *Eurogamer.*

Williams, K., & Nida, S. (2011). Ostracism: Consequences and coping. *Current Directions in Psychological Science, 20*(2), 71–75.

Williams, D., Kuipers, A., Mukai, C., & Thirsk, R. (2009). Acclimation during space flight: Effects on human physiology. *Canadian Medical Assocation Journal., 180*(13), 1317–1323.

Wilson, E. O. (1975). *Sociobiology: The new synthesis*. Boston: Belknap Press.

Wilson, E. O. (1984). *Biophilia*. Cambridge: Harvard University Press.

Wilson, J., & Kelling, G. (1982). Broken windows: The police and neighborhood safety. *Atlantic Monthly, 249,* 29–38.

Wilson, M., Boesch, C., Fruth, B., Furuichi, T., Gilby, I., Hashimoto, C., et al. (2014). Lethal aggression in Pan is better explained by adaptive strategies than human impacts. *Nature, 513,* 414–417.

Wolfe, T. (1979). *The right stuff.* New York: Farrar Straus & Giroux.

Wolff, H., Silberstein, S., Lipton, R., & Dalessio, D. (2001). *Wolff's headache and other head pain.* New York, USA: Oxford University Press.

World Health Organization. http://www.who.int/mental_health/prevention/suicide/suicideprevent/en/.

World Health Organization. http://www.who.int/mental_health/en/.

Yamane, K., Kawashima, M., Fujishige, N., & Yoshida, M. (2004). Effects of interior horticultural activities with potted plants on human physiological and emotional status. *Acta Horticulturae, 639,* 37–43.

Zhang, J., Yao, W., & Hashimoto, K. (2016). Brain-derived neurotrophic factor (BDNF)-TrkB signaling in inflammation-related depression and potential therapeutic targets. *Current Neuropharmacology, 14*(7), 721–731.

Zimmerman, M. (1989). The nervous system in the context of information theory. In R. Schmidt & G. Thews (Eds.), *Human physiology* (pp. 166–173). Berlin: Springer.

Zubek, P. (1969). *Sensory deprivation: 15 years of research.* New York: Meredith.

Chapter 2
Program Planning for a Mars Hardship Post: Social, Psychological, and Spiritual Services

Margaret Boone Rappaport and Christopher Corbally

Abstract Human services planning for crews who go to Mars is in its earliest phase, but the modalities for service delivery are well worth anticipating because they could involve some of the first innovations that merge physical, biological, and digital capacities on the new planet. This chapter examines the constraints of the planet Mars, itself, on all humans. It anticipates how "exogenous stressors" (gravity, atmosphere, radiation, light, and dust) might affect the psychological, social, and cultural capacities and conflicts of the earliest crews. Several types of service modalities are explored: hibernation; medication management; traditional psychotherapy; long-distance modalities to approximate the "talking cure"; digitized, keyed psychology workbooks; robotics, or service delivery without verbal feedback; AI technologies, or service delivery with verbal feedback; problem-solving at the cultural level; digital bulletin boards to substitute for journaling; and a special section on the issue of privacy. The chapter rounds out with a section, on types of spiritual services that may be beneficial even for those who profess no faith. We focus on "sacred space," and on death and burial ritual on Mars. Factors explored here reflect the backgrounds of the two authors: one, a cultural anthropologist and biologist, and the other, an astronomer who is also a Catholic priest.

> March 29, 1700…And what great news and how rare is that which your Reverence imparts to me…that it is possible to pass overland to California, news truly the greatest, if it is verified, but which, although desired so long, it has never been possible to confirm… October seventh… we had seen the lands pertaining to California, without any sea between and separating those lands from it (Kino 1919, orig. 1683–1711).

M. B. Rappaport (✉)
The Human Sentience Project LLC, 400 E. Deer's Rest Pl, Tucson, AZ 85704, USA
e-mail: msbrappaport@aol.com

C. Corbally
Vatican Observatory, U. Arizona, Tucson, AZ, USA

K. Szocik (ed.), *The Human Factor in a Mission to Mars*,
Space and Society, https://doi.org/10.1007/978-3-030-02059-0_2

2.1 Introduction

The human enterprise on Earth has long had Mars in its sights, and now the time is nearing when our species will set foot on that planet. This adventure joins many others in the past, including the discovery of a land route to California by Jesuit explorer and scientist Fr. Eusebio Kino in 1700. While the Native Americans who roamed these lands knew the route to "California" well, Europeans did not. There was a sense of adventure in Kino's travels and that sense will not change in our first Martian expeditions. The species remains the same, and our longing to explore will carry us, later still, to other planets and stars.

This chapter addresses human needs and how to meet them in the very first settlements on Mars. They will not be easy places for humans to live. We anticipate some of the problems, some of the services needed to solve the problems, and how these services could possibly be delivered so far away from our home planet. We surely have not anticipated everything, but we hope we have injected a sense of realism about this task by fully acknowledging the complexity of human social and cultural life. Solutions cannot be sought within a simple matrix of needs and capacities, but anticipated with the full appreciation of the daunting, but inevitable, nature of this human adventure.

2.2 A Mars Hardship Post

The term "hardship post" is borrowed from a specific type of program and mission crew. It is a term used by the United States Department of State when assigning foreign service personnel to countries and regions that impose unusual dangers, inconveniences, or health challenges. Here, we borrow "hardship post" for, in some ways, a similar type of personnel that will all be exquisitely selected, thoroughly trained for years, and instilled with a sense of mission dedication. It is expected that early Mars crew members will need all this background and some luck, too. The hardships that the first crews to Mars will endure will be like none experienced on Earth in Yemen, Bangladesh, or Somalia. There will be no wars and no terrorists, one hopes, at least not at first. The dangers will be fewer in some ways, but far, far greater, in other ways.

Our purpose here is to address the types of stressors and the modalities for human services delivery to alleviate them for the first Mars crews. We are especially interested in Martian crews who begin the change to a human "community"—encompassed, self-aware, self-serving, and long term. This transition has been described in several ways, for example, the transition of early exploratory crews to settlements where inhabitants are aware that they will probably never return to Earth. Acceptance of Mars as "home" encompasses living out all of life's joys and vicissitudes on a different planet. That point of view should be substantially different from the perspective of early crews. It is not altogether clear when this transition might occur, since it may be unexpected. Humans who settle on Mars with the expectation that they will return to Earth may find themselves unable to do so, or not wanting to do so. Until that time, and beyond that time, services will be needed to help humans adjust.

We first address the exogenous stressors that characterize the planet Mars, and which together impose enormous burdens on humans. Their combination provides perhaps the most pressure to the human psyche, the human body, and the human spirit. One of the Mars stressors we describe will be difficult to manage, but all of them at the same time introduce the possibility of multiple systems' failures, or, different crew members with different types of issues. Instead of focusing on emergencies, we focus on routine types of human difficulties—psychological, social, and spiritual—and suggest modalities for managing them.

Natural, human estimations of risk will not be blind to the dangers of the Martian planet, and the culture of the early Mars crews is sure to change in order to provide some softening of the perceived risks. For hundreds of thousands of years, humans have been constructing rationales, explanations, and myths that buffet themselves from the risks of both environmental dangers and the challenges of their own bodies and minds. This is human cultural capacity at work, and we will address the ways that culture, in combination with human services, may help to create hope and faith in the future for Martian crews and eventually, communities. Convictions about the future will come to rely increasingly on the Martians, themselves, and less on their mission control support on Earth. The first crews and communities will need services that can help them problem-solve their own unmet psychological, social, and spiritual needs. These are not signs of weakness, but ultimately, signs of human progress in establishing the beginnings of a lasting community.

In addressing different types of modalities for service delivery, we ask: How can the usual, earthbound services be delivered when the roundtrip communication time lapse is up to 40 min? There are ways. How can artificial intelligence, robotic technologies, genetically engineered modifications, and bio-enhancements—in addition to more routine forms of therapy and self-help—all help humans cope with their environment? Technologies are rapidly emerging to provide options for supplementing service delivery on Mars.

We address an area rarely considered for Mars crews—the provision of spiritual services (*cf* Szocik 2016; and Oviedo in this volume, on "Raising the Right Questions" in planning for religion on Mars). Here, we ask how spiritual services can be delivered when the cultures and the faiths of the crew are varied. However, this issue may soon fall away as crew members find their own sources of spiritual enhancement on a new planet, and by themselves engineer a new cultural synthesis for a new planet.

Belief systems will begin to change almost immediately. Even during the long voyage out to Mars, which is well over 48 million miles farther from the Sun than Earth is, crew members may develop an awareness of moving away from the Sun's heat and warmth. They may mourn the loss of Earth and its bright familiar star in Earth's sky, just as they might grieve a person, as we note in our final section. Their sense of danger and isolation from familiar forms of support may grow with their awareness of their increasing distance from Earth and the Sun, along with a cognitive shift in awareness of Earth's fragility called "the overview effect" (NASA 2011, 29). Since the times of *Homo erectus*, before our species evolved, human groups defined comfort, food, and social life according to the warmth and light of the campfire. Our species remains acutely aware of the antithesis—darkness, cold, and solitude. Those qualities define space.

The culture of the crew will change, coalesce, and produce a new synthesis that is unique to the psychological and spiritual supports they need. When they arrive, we speculate on which aspects of the new planet will call forth conflict and irritation, but we also ask which aspects will call forth human awe and wonder. Which locations in their environment will offer space for spiritual reflection? The crews will find them soon, along with changed values based on altered equations about life and death, comfortable and uncomfortable, human and inhuman, and right and wrong. Those values will suit a new planet, not the old one. Support teams on Earth should be prepared for these changes in values and perceptions of risk.

As a Martian culture evolves, so will the population genome change—at least in small ways, as we have seen in the twin study of the Kellys, with one astronaut on the International Space Station for an extended period (NASA 2018). Recognition of the real potential for rapid human genome change is gaining a foothold and has been for some time (Hawks et al. 2007; Cochran and Harpending 2010; Solomon 2016; Seto et al. 2017). On Earth, rapid human genome change is most often seen as a result of isolation, but it is also possible on Mars, where Founder Effects, genetic drift, and genome change may be obvious because the population is small. It is likely to remain so for some time. On Earth, genome change will be more difficult to observe because the human population is so large. Still, natural selection is most effective in large populations, while genetic drift works better in small populations, as on Mars.

If the Martian genome changes, we ask: Will the crew think of themselves as "human"? If not, what? These changes in cosmology and self-perception have been seen time and again in human communities on Earth among people who move from rural to urban environments. They see themselves differently. They call themselves by some other name. What will the early Martians call themselves?

Two additional factors may impact all types of program planning for manned Mars missions: Mars versus lunar settlement; and commercial versus nation-state sponsorship.

By the time humans go to Mars, there will be a knowledge base on off-world settlement on Earth's Moon. The order of settlement has appeared different lately, while Mars takes temporary prominence. However, it is more likely that human settlements will emerge first on the Moon. We will learn from that experience and be able to plan more cleverly for the challenges faced by Mars crews. We have none of that knowledge base now, so our extrapolations necessarily "jump over" lunar settlement, to speculate about Mars.

A second major unknown is the source of sponsorship for the first Mars crews. Few writers have addressed the monetary compensations that might encourage crew to join a Mars mission. Like the State Department's foreign service personnel, financial incentives may well be incorporated into mission planning, especially if privately funded Mars missions gain a foothold. Today, they are still in the hypothetical stage, and "payoffs" for crew who undertake nation-sponsored missions remain adventure, exploration, discovery, and a sense of duty. Over and above these personal motivations, the goals of the mission sponsors will be important to an emerging Martian culture. If the initial incentives are commercial and aimed at exploiting the natural resources of Mars, the culture will take on those values. At stake in mission goals and the incentives of mission sponsorship is the nature of the society that is built on Mars: Its economics, politics, and ultimately its relationship with its home world, Earth.

2.3 Exogenous Stressors

Human experiences on Mars will be shaped strongly by the nature of the planet, itself, which is smaller, colder, and biologically barren, at least as far as we know. The protections afforded by the warmth of Earth's Sun, vegetation, and surface water will be gone. The dangers of predation will also be gone, unless stress on human crews becomes overwhelming and they fragment socially and turn against each other. It is well worth anticipating both the psychological and social stressors that would render an early Martian community dysfunctional, and which services would keep it running smoothly. Little social and cultural research has been done in those areas.

It is as if we assume that the excitement, accomplishment, and human awe at a different and very beautiful new environment will be enough to keep social and cultural life at least stable, if not upbeat. However, the sheer number stressors and their basic quality may quickly overwhelm any sense of adventure. In the type of extreme environment posed by Mars, it is worth remembering how humans have solved problems in the past—and that is, in groups. Humans are not only social, but they descend from Primates who all lived in troops. Solutions are group solutions. Furthermore, after *Homo erectus* emerged, some of them took up a nomadic way of life that enabled them to leave Africa. *Homo sapiens* only augmented this tendency to roam, and we colonized Earth in fairly short order—given the fact we were walking and had only crude tools to defend ourselves. In all the very new habitats humans conquered, one after another were tamed by human reliance on social organization and cultural creativity to keep hope alive in the face of multiple life-threatening hardships. Still, we never conquered a poisonous atmosphere where the only place we could breathe was in sealed habitations. Confinement may be a great problem for Martian crews, but they will solve it—with short journeys when they can, by walking in protective suits for short distances, and perhaps with inventive technologies that help to bring Mars into their habitations. Impey has suggested "large format, flexible, programmable, liquid crystal displays that 'map' the exterior view onto an interior wall with high fidelity" (2018). To this, we add virtual reality programs and broadcasts, which can also be useful in training, exploration, and rescue.

"Stressors" have been summarized for humans in space, for example, Kanas (2015) lists the following: periods of high acceleration, microgravity ("weightlessness"), ionizing radiation, meteoroid impacts, light/dark cycles, vibration, ambient noise, low and high temperature, diminished lighting, and no air or poor air. In addition, Kanas summarizes "psychosocial stressors" in space journeys, for example, isolation and confinement, life-threatening danger, periods of monotony versus high workload, reactions of depression or elation to specific time periods of a space journey, personality and crewmember selection issues, too much free time, increased autonomy, and dependence on machines and local resources. Kanas labels as "interpersonal stressors" the following: group size, limited social contact and novelty, group heterogeneity, common language, cultural differences, governance structure, and leadership. Cockell (2016) explores the stresses of different government forms in a human space base.

Some of these stressors during space travel will transfer to Mars. Nevertheless, the multiplicity of interactions among even these very short lists of social and cultural factors suggest that program planning must eventually shift to the service modalities that can address a far, far wider range of psychological, social, and spiritual issues than those examined, so far. Human social life is a great deal more complicated than anything anticipated, to date, for what we term a "Martian hardship post." Isolated military bases and hospitals in wartime, on Earth, may provide better social and cultural models for the initial Martian settlements.

This approach is not a pessimistic one, but realistic and in some ways, optimistic. Human hardship posts of a variety of types—from isolated and confined Antarctic science crews, to prisoner-of-war communities, to lone prairie households in the 1930s American dustbowl—have demonstrated how resilient, creative, and strong our species is when we confront daily, life-threatening dangers and maddening environmental factors. The ability of humans to create solutions when they work together with a shared purpose and a common belief in the future is remarkable, and will be so, on Mars, even if the first settlements are failures. They may be, but that will not dampen our enthusiasm for trying again and again, until humans get it right. Let us review the aspects of Mars that may produce the most pressures and the greatest stress among the first crews.

We include an analysis of stressors within a context of sociocultural change. This is the missing context, to date. Lists of stressors "jump over" the higher cognitive capacities on which humans on Mars will have to rely, to survive. With a longer and broader perspective, we may discover that identifying "The Right Stuff" is a great deal more complicated than first anticipated for crew of early, experimental space vehicles. Human cognitive capacities have varying ranges of adaptive expression, which result, in part, from our species' developmental plasticity. Add to this inherent variation an even wider range of possibilities with genetic, physiological, and biophysical enhancements, and program planners may be able to target likely candidates with increasing chances of success. Most important will be to anticipate problems and contemplate a full range of enhancements—including human services delivered on Mars, itself.

Gravity. The gravitational pull of the planet Mars is weaker than Earth's because of its smaller mass. At only 38% of Earth's gravity, the human body will feel lighter and move differently on the planet, irrespective of protective clothing, which will also impede locomotion. We note that while the "Psychology of Space Exploration" has been reviewed in "Historical Perspective" (NASA 2011), it is our hope that the challenges facing humans on Mars will also be seen also from the perspective of prehistory. After all, bipedalism (partial or full) is the method of locomotion that distinguishes members of our genus Homo. It emerged over millions of years of evolution on Earth, and on Mars it will change, primarily as an adjustment to lower gravity. The muscles and bones of the legs and the torso have evolved to accommodate the bipedal gait of humans on Earth. It is extremely efficient and able to propel a hunter on a long trek that eventually outpaces and exhausts a big cat or a horse.

All of that will change on Mars. Some muscles that were barely used may become newly important, and the large gluteus maximus (the largest muscle in the human body), which advances humans in the hunt (or in a marathon) so efficiently on Earth, may become less important. Ackermann and van den Bogert (2012), in anticipating some of these changes, have conducted simulations that suggest "skipping" will be the preferred human adjustment in the lower gravity environments of the Moon and Mars. They note correctly that research prior to theirs focused primarily on walking and running, which are preferred in Earth's gravity. They find that skipping is more efficient and less fatiguing than walking or running in lower gravity.

The social consequences of skipping may be profound, for example, because it makes "walking together" more difficult. Indeed, slight modifications to the skip may be necessary for confined spaces on Mars to reduce the distance covered, since the skip propels the human body both upward and forward with less control than the normal walking gait. Weighted booties are an option that could provide humans with finer locomotor control.

There are other unknown consequences of a lower gravity on Mars, including physiological processes, circulation, digestion, and reproduction (which is likely to be prohibited until research is conducted on its safety) (Szocik et al. 2018). Bone density both in space and on the Moon and Mars will be reduced, and humans will be in danger of bone breakage. Options for prevention and treatment run the gamut from exercise to drugs to genetic engineering and genome change (Szocik et al. in press). The psychosocial stressors emerging because bodies are changed—stronger in some ways, but weaker in other ways—will have profound effects on mood, interpersonal interactions, and workload. Who is "strong" and who is "weak" among the crew may change in definition, with alterations in work roles, leadership roles, and self-esteem.

Atmosphere. The atmosphere on Mars is very different from Earth's in two major respects. First, it is extremely thin, at less than one percent of Earth's atmosphere. Second, it is composed primarily of carbon dioxide at about 95%, nitrogen at 3%, and argon at 1.6%, plus some trace gases and water vapor. Small amounts of methane have been discovered, and they may result from either organic or inorganic processes. The Martian atmosphere is also filled with iron oxide dust, which gives the sky a beige to pink to rust color from the surface. We address dust below as a separate factor.

The implications are fundamental for humans. It is a poisonous atmosphere, so it will be impossible to survive without reliable technologies that provide, for now, a completely manufactured and carefully controlled air to breathe. There are efforts to produce conversion equipment that takes the carbon dioxide from the Martian atmosphere and uses it to produce oxygen for humans to breathe, and, as fuel for rockets. Michael Hecht, at MIT, has experimented with developing an instrument to use carbon dioxide (recall, each molecule has one carbon and two oxygen atoms) to produce gaseous, diatomic oxygen (Gaudin 2014). Hecht's will surely not be the last of such efforts because of the potential profitability of mining oxygen fuel in the future. Mars has the potential to be an important break-in-bulk point for space ships to the outer planets and as a supply station for oxygen fuel and other consumables. The Martian atmosphere is very thin at 600 pascals, compared to Earth's 101,300 pascals. Nevertheless, on Mars the atmosphere extends farther from the planet because of its lower gravity, which does not hold the atmosphere as close to the planet and because solar winds are constantly reducing the density of the upper atmosphere. Mars does not have magnetic poles. The planet lost its magnetosphere billions of years ago, so it lacks the protective "shell" to prevent loss of its atmosphere to the same solar winds, as well as to prevent some radiation.

Radiation. There are profound consequences of the thin Martian atmosphere for living, working, and greenhouse farming on the surface of the planet. For humans, the Martian atmosphere fails to form a safe "cocoon" like that of Earth, preventing most direct solar radiation. The consequences of this lack of protection for the human psyche and for personal interactions are unknown, but awareness of a thin, poisonous atmosphere and constantly incoming radiation will remind the first crews of the hostility of the environment, and the need for a technically controlled oxygen supply and protective equipment and covering. Safety protocols will be practiced and re-practiced to ensure every human's ability to safely explore the surface, and to rescue crew mates when protections are breached. The constant readiness for an emergency may render humans hypervigilant. Technically, hypervigilance can be seen as either a rational reaction to the real dangers of the environment, or when unwarranted, as neurotic and sometimes causing extreme anxiety.

We noted above that cultural capacity works to buffet humans from constant anxiety about environmental dangers. Culture provides explanations for dangers even when the science is unknown, makes danger predictable (even when it is not), and defines "safe" circumstances even when they do not exist—all to reduce human anxiety. Furthermore, we are discovering that certain human genome components make some humans more sensitive than others, and in different ways (Aron et al. 2012; Acevedo et al. 2014; Todd et al. 2015). It is worth noting that these genes exist only in proportions of populations tested, not in all. We see hypersensitivity and hypervigilance as counterproductive in all members of a group. Again, human solutions are group solutions, and it may have been useful during our evolution to have some individuals who were particularly attuned to danger but not all. What this says for selection of Mars crews is debatable: Would it be better to have crew in which everyone is less sensitive, or in which a few were quite sensitive and vigilant? Now that we can test for these genes, it makes their use in selection a realistic option.

The experience of walking or riding on the Martian landscape, and fully experiencing it firsthand, may have features akin to deep sea diving because of the need for protection from radiation and breathing equipment. At the same time, it may elicit awe and wonder, just as astronauts have been struck by the whole world view of the Earth from space. It may elicit some innate or acquired tendency for claustrophobia due to confinement, or agoraphobia in response to vast Martian distances, or both. Navy pilots and SEALs are routinely "pushed to the limit" and tested for tendencies for these types of anxiety reactions. However, testing cannot be foolproof. When it comes to landscapes, a visual scene that is one person's anxiety trigger is another person's invigorating view. As a species, humans are acutely tuned to visual landscape features, and their colors and shapes. Dutton claims that a "fundamental attraction to certain types of landscapes is not socially constructed but is present in human nature" (2009, 18). Cockell (2002) has warned of the monotony of the red color of the Martian landscape, and we would add, a disquiet at the absence of blues and greens, and streams and trees. Artistic forms of recreation for the crew may be particularly useful in their comprehending the new and different landscape. In some locations, Mars has an exaggerated topography for crew accustomed to Earth, for example, the depth and breadth of the Valles Marineris and the height of Olympus Mons. This is not to say that humans cannot accommodate to visual inspection of the surface—they will—only to say that there may be challenges, and suits hinder reassuring human-to-human interaction. Early model mobile robots to accompany explorers may represent both safety and company, at once, like dogs once did, for earlier members of our genus Homo.

The extent to which the implications of these scenarios have been thoroughly considered by program planners is unknown. It may be that the only surroundings where humans can truly relax are inside domes, bunkers, and insulated habitations. On Mars, the most obvious protective material is the regolith, itself, so it will not be unusual to find the first crews taking with them tractors, backhoes, and other heavy equipment. They will all be expert at using them, too. It is likely that the first Martian crews will grow accustomed to the Martian surface more by moving it around, than by transporting themselves on foot or in small ground cars around the planet. Design of redundant air supply equipment for tractors and cars must eventually take a high priority.

None of these problems will hinder humans from exploring the surface of Mars, because that is what they have evolved to do. Humans were exploring Earth's Old World land masses as an earlier form, *Homo erectus*, beginning around one million years ago, before our species, *Homo sapiens*, even appeared about 300,000 years ago (Hublin et al. 2017).

It is worth contemplating that the assignment of activities on the surface, their scheduling, planning, and implementation, may be one of the earliest sources of interpersonal tension and therefore a management challenge. After being confined in a space vehicle for a trip that could take over a year, humans will want to explore the surface for scientific and personal reasons behind the urge to explore. They will see exploration as a right and a duty, no matter how well trained to the contrary. Selection of crew to conduct reconnaissance may raise some of the earliest tensions. Competition may be keen to "get away", map, sample, and experiment with the few available vehicles.

Light. The color of light from the Sun is yellowish, and the spectra of light are the same on Mars as on Earth. The important difference is "irradiance" or intensity of light striking the two planets. Mars receives less than half of the light we receive on Earth but the proportion available for photosynthesis differs substantially. A few figures will show the difference. On the surface of the Earth, with the Sun overhead at noon, irradiance is about 1000 W/m^2. At noon on Mars, under the same conditions, irradiance is about 590 W/m^2. Why the discrepancy? Why does Mars have more than half the watts when it receives less than half the light? The difference is due to Earth's reflection of light—its atmosphere, surface, and clouds. Mars has a much thinner atmosphere and no clouds. Therefore, there is both good and bad news for the future of greenhouse agriculture on Mars.

Because the opaque Martian regolith is the most readily available substance to shield humans whenever they are not in protective enclosures or suits, crews are very likely going to be exposed to very little natural sunlight, and this may pose problems for vitamin D production, bone health, and maintaining mood. Whatever small "windows" to a Martian sunset exist in the first habitations, they may become either valued or repugnant locations, when crew realizes how small the solar disk is when viewed from Mars. We wonder whether viewing a Martian sunset will be reassuring or unimportant or disturbing. That may depend on exactly how much Martian sunlight can be put to work to support greenhouse agriculture.

Existing designs of inflatable bubble-like buildings and cheery greenhouses with wide widows may be unrealistic. The best design could be placement of the crew habitation underneath the greenhouse, where it can be shielded by the regolith. That said, unless materials scientists can design an extremely hardy, transparent yet filtering material to capture the available rays of the Sun on Mars, and, simultaneously protect Earth-evolved plants from irradiation, the plants may receive too much direct radiation. The greenhouses on Mars will also be subject to breach by periodic small meteorites. There is no suitable translucent material yet that allows enough of the sunlight reaching Mars to support photosynthesis, and that also protects plants. That is a very tall order, but not impossible, especially as nanotechnologies advance. More damaging still to Earth plants are the dryness of Mars and the very cold night temperatures. Breaching a greenhouse on Mars with a sizable meteorite will quickly produce freeze-dried vegetables, so the best designs include partitioned greenhouse units that can be quickly isolated, if necessary.

After human protection from radiation and a poisonous atmosphere, the goal of ensuring a sufficient food supply will be primary. The qualities of the transparent materials for greenhouses will ideally include options to change filtration when, for example, a Martian dust storm lasts for months and light is even further reduced. Greenhouses and solar panels have some of the same issues on Mars—protection from meteorites and fine-tuning for available light spectra. It will be important to establish early in the course of Martian settlement exactly how much plant material can be sustained by a weaker Martian sun and how much artificial light will be needed to supplement. It may well be easier to match optimal light for photosynthesis using artificial light. Much of the research can be done on Earth, but not all. The bioengineering of plant foods may be one of our earliest priorities, to change the ability of plants to use more wavelengths of sunlight, and to withstand more radiation. After all, both humans and plants evolved on an extremely different planet Earth. Changing their genomes to accommodate Mars is an increasingly likely option for both, although the ethical implications are very different.

Solar panel research to withstand the constant incursion of Martian dust will be a challenge, and equally problematical will be panels that can receive light but protect the "sandwich of materials" from micrometeorites striking the planet. Both dust and meteorites will challenge the ability of Martian crew to obtain and maintain sufficient energy to run the electronic equipment they bring. Battery technology will have the benefit of being able to be buried in the regolith to protect it, but sufficient storage capacity will be a problem for the foreseeable future. All of the equations for the use of solar energy will change, and the crew will, by necessity, change their notions of what is, and is not, reasonable energy expenditure. The design, construction, and testing of energy production and storage systems will involve humans on Mars for many hours of a single mission and for many years for mission planners.

In the meantime, we ask: How do humans fare in much reduced solar light? How do they live in an essentially red environment? How do they maintain work routines, social activities, health, and mood in severely reduced light, as during a Martian dust storm, and at the same time, in cramped spaces? The slightly longer Martian day may actually be a better match for the human body rhythm, although these vary widely, as does tolerance to varying amounts of sleep. How are moods regulated in the face of the most exciting color for humans—red, the complement of green, the most soothing, and nowhere to be found on Mars except in a greenhouse.

It may be that the very definitions and understanding of "social" and "cultural" will have to change, and if that is true, what does it mean? "Personal space" may come to mean something far different in Martian habitations. "Social time" may become much more strictly defined, or alternatively, much looser. Cultural rules for when to intrude in someone else's space and on someone else's "alone time" will change. Privacy will be guarded, or perhaps abandoned altogether. Drugging and hoarding (of food, games, or lipstick), or other obsessive–compulsive routines may arise to aid the crew in tolerating these new conditions. Curtains may be drawn and doors closed—or left always open. As in some settlements in Antarctica, individuals may install hammocks in the greenhouse, just to maintain mood. Someone may jealously guard the best view of the Martian sunset. Every single crew member will be trained to be on the lookout for the signs of alienation, despair, depression, excessive withdrawal or euphoria, and mood swings. They will also be trained to report it, perhaps daily and at least weekly, because the consequences can be deadly, along with the destruction of finely tuned and very expensive scientific equipment. If mood is allowed to decline too far, for too many, over too long a time, the settlement will fail.

While these developments sound dire, they could be wrong, or they could be understated. When we take a look at other higher Primates, like the bonobos, who have a less aggressive and possessive style when compared to chimpanzees, we find troop animals that socialize constantly, crawl and fall over each other without much fuss, and are quite docile and noncompetitive. They are also hypersexual—will humans be so? Bonobos, chimpanzees, and humans are very different species, although they arose from the same ancient stock of Miocene apes. How much of the other species' potential remains in the human genome to exploit and to use creatively? The genomic science to unravel answers to that question remains in its infancy, although it is growing fast.

Dust. We address the issue of dust because it crosscuts the exogenous stressors mentioned above. Humans have only encountered dust like Martian dust on the Moon. It is finer than anything on Earth because the biome, surface water, and geological upheavals are constantly churning organic and inorganic material, and Martian-like dust has no chance to settle or accumulate. Martian dust will penetrate each and every seal devised by humans from Earth. Nothing will be completely impenetrable except perhaps for a protective suit. Each time a suit is worn on the surface, it will take some Martian dust inside human habitations and greenhouses. Martian dust will cut the sunlight during a dust storm. It will swirl and winds will blow it around for weeks or months, but there will be no howl to the wind because the atmosphere of carbon dioxide is so thin. The greatest sound might be micrometeorites tapping on the greenhouses or human habitations. Furthermore, dust storms on Mars will include dust devils that have electrical properties that can short-circuit expensive equipment.

In light of these predictions, it is reasonable to assume that dust abatement will arise as an entirely new inventory of techniques and appliances. To date, it has not assumed the expected priority that we suggest is reasonable. The concern about dust raises important questions about human health. Unlike dust on Earth, there is almost no water vapor, microbes, spores, and other organics that make Earth dust clump and fall down and away. It can be dislodged from the surface of Mars with a single step of a single boot. It can be carried aloft to circle for months.

How will humans react to dust that is so intrusive? How will they manage dust abatement efforts? Will they feel as if they "can never get clean"? Will it be just a simple annoyance that they learn to live with? Will it be maddening or, in the end, normal? These questions are important because Martian dust cannot be collected in a handy vacuum cleaner, removed, and forgotten. It is everywhere, and it will be everywhere that human crew live and go. Will handling dust change the cultural value of "dust-free"? Will new definitions of "clean" arise, and what are the consequences for health and hygiene? Rock dust (talcum powder) has been found to be connected to ovarian cancer. Will Martian dust increase the chances of lung cancer in humans?

Let us emphasize that in the preceding sections, we have focused on exogenous factors, not the individual personality characteristics of likely crew, nor the issues that people in specific occupations might have. We have mentioned, but not concentrated on supervision and management, which crosscut occupational categories and personality types.

Nevertheless, it is obvious that the types of problems encountered by early Mars crews cannot be reduced to shortlists, so in the following section we suggest ways in which psychological, social, and spiritual services can be delivered to help solve a broad array of problems. Hopefully, they can address the panoply of potential personal, interpersonal, and community problems. The entirety of these problems reflects the complexity of human social and cultural life. We know something about how "small societies" change and we have some general guidelines for how cultures change, but human complexity is not to be minimized, and we suggest service modalities that can be broadly targeted or modified for individual cases.

2.4 Modalities for Service Delivery for Early Mars Crews

Long-term living in a threatening environment will inevitably produce stress and anxiety in most people. The longer the time away from Earth, the greater will be this stress (Kanas 2014, 103). Living on Mars will call for special services to overcome or mitigate stress. Indeed, a "portfolio of models of delivery will be needed" (Kazdin and Blasé 2011). Let us look at a potential portfolio of service types that can address a combination of psychological, social, and spiritual issues. Our selections necessarily take advantage of some of the latest information technology capabilities, including some that are not yet fully developed.

Hibernation. One method for controlling the stress of spaceflight is for humans to remain in a state of induced unconsciousness during the long voyage to Mars. The terms hibernation, suspended animation, stasis, and recently, torpor (Bradford et al. 2014) have been used for this condition. Granted, hibernation is not usually considered a form of service delivery, although perhaps it should be, because it manages symptoms of *stress* by preventing their development. Stasis could also be induced if one of the crew becomes injured or ill, and no medical help is available during the flight to Mars or after landing (Szocik et al. in press). Biologist Martin Braddock succinctly summarizes the physiological basis for the approach: "The potential for human beings to be in a state of hibernation or stasis has long been recognized from the animal kingdom where animals, including all type of mammal have an innate body clock which enables them to safely and spontaneously enter and leave hibernation" (2018, 2–3). He points to widely used "controlled hypothermia" in surgery and in the case of brain trauma, which has been used for clinical purposes for up to two weeks (Schaffer et al. 2016). If used for longer periods, it is possible that atrophy of bones and muscles will become an issue. Some researchers feel suppression of metabolism may be the key, and there is substantial, ongoing research, for example, on the Arctic ground squirrel who hibernates for two-thirds of the year with only minor bone and muscle loss. There is a broad and growing literature on the physiology of this squirrel and other hibernating mammals. Kelly Drew, a pharmacologist with the University of Alaska Fairbanks, and Alejandro Rabinstein, a neuroscientist at the Mayo Clinic, are among researchers working on hibernation with spaceflight in mind. One of the most problematical issues related to stasis regimens is the absence of an automated technology to safely take humans in and out of hibernation (Braddock 2018, 1).

Once humans have exited stasis, there are indeed a variety of "services" that require planning, including physical and mental rehabilitation, if necessary. There are social and psychological services that range from accessing family bulletin boards to learn news that occurred while the crews were in hibernation, to bereavement counseling in case someone in a crew's family has died. Spiritual meditation or services can be organized for the reawakened crew along the lines we mention toward the end of this chapter.

Medication Management. The crew of all early Mars expeditions and early "hardship posts" will surely include at least one medical officer (physician), perhaps more. Use of psychoactive drugs to alleviate anxiety and depression have been addressed before, and so we shall only add to those interventions, changes to the genome that may eventually be considered to produce crew members who can better tolerate the dangerous conditions of an early voyage and settlement. At the present, there is understandable antipathy to drugging or pre-mission genetic alteration of an entire crew, but given the obstacles that early crews face, these remain unhappy options. Ideally, crew will be selected who can manage long-term stress, but then, we have no experience with years-long assignments in space. If it is a choice between illicit selfmedication and supervised medication, the latter is clearly the best choice. Preflight genetic alteration of humans to control their mood is a science that is still in its infancy, and the ethical problems that attend this possibility are substantial and may be insurmountable.

Traditional Psychotherapy. In addition to the medical officer(s), there may be one or two crew who have some experience in counseling. All of these individuals would be familiar with the concept of a "talking cure" where regular meetings in private are used to diagnose and treat psychological problems that are preventing crew from assuming their work duties. They will be familiar with the issues that go along with transference between therapist and patient. Still, it is hard to imagine any circumstance more difficult for traditional psychotherapy than an early hardship post on Mars. A monotonous voyage to Mars could afford the time and privacy to engage in therapy like this, and in some cases, it might be considered desirable. It is more likely that crew selection would be sufficiently fine-tuned that none of the early crews are in great need for intensive psychotherapy, and yet we do not know this for certain. Once the crew arrives on Mars, it is very likely that staying alive and establishing a settlement will consume the energies of all the crew, and only the most extreme psychological distress will receive attention.

To date, even fictional accounts of Martian settlers have failed to address the issue of care for the caregiver. Ideally, the physician(s) and/or counselor(s) for Martian crew would be routinely be trained in the dangers inherent to taking on others' psychological burdens—including dangers to physical health and susceptibility to drug abuse. Therapists on Mars may find that a "therapist's therapist" or "supervising therapist" is a very good safety precaution, and that one could be available via correspondence with another therapist on Earth. In hardship posts and hospitals in wartime, it is extremely difficult for any therapist to maintain distance from his or her work. In this particular case, a healthy perspective on a Martian crew that is very rapidly changing, socially and culturally, in front of their eyes, will be even harder to maintain.

Long-Distance Modalities to Approximate the "Talking Cure." Once on the planet, there may be a need for some type of intervention to substitute for therapies, other than medication. Talking, and later writing one's resulting reflections, have long been therapeutic channels open to humans. We can readily assign the "talkers" and the "listeners" among our acquaintances, and those who combine both skills are especially valuable friends. In the first Mars crew, the pool will be limited to one or two who are trained in psychotherapy among their other competencies. Other resources will be needed from both on and off the planet.

One resource is to use a group as the therapist (Forsyth 2015). The stressors outlined in the previous section will be held in common by the crew. There will be much to share concerning the feelings that they have in relation to these stressors, and how to develop solutions. A crew member with some training in group therapy will be helpful since the interaction within the group has to be immediate. However, any leader will be helpful in guiding community meetings. Use of a Martian group leader on Earth would be hindered by the time it takes for communication signals to travel to Earth and back. An added benefit of group meetings on stressors will be the identification of individual crew who are having great difficulty.

Another modality that uses the group in a potentially anonymous way is an electronic chat room that addresses stressors as they occur. Crew members would post their thoughts, concerns, fears, and successes anonymously in a digital, shared space. It would be a type of social media workspace to hash out problems common to some or many. This could be seen as a type of service delivery by the group, itself.

A modality for service delivery that can withstand the communication delay is an exchange based on journaling. This is the practice of writing down the difficult and pleasant parts of one's life, and its tradition dates back at least as far as tenth-century Japan. Self-reflection on the joys and woes of the day is also evident in the mental "examination of consciousness," an exercise introduced by Ignatius of Loyola, the founder of the Society of Jesus (Jesuits). Journaling has certainly been found to be effective, but focusing on factual understandings as well as emotional reactions is important for its success (Ullrich and Lutgendorf 2002). A journaling correspondent on Earth could be chosen from a wider pool of people, although a close friend might be found among the crew.

There are obstacles to journaling on Mars. The physical writing, with pen and paper, seems important to journaling for many people. Among Mars crew, laptops and tablets will have to suffice, although touch screens may help to solve this problem. It is true that some people simply do not take to journaling, whatever encouragement they receive. Snyder (2013) has reviewed the practice of journaling and suggested alternatives, such as art. Mars mission planners could incorporate forms of art and even an "art area" in the Mars living facility, with the goal of self-expression. This type of "psychological self-service" may sound elementary now, but if it prevents destructive forms of emotional expression, it will be well worth it.

Engineers, military officers, and pilots will dominate the profiles of Mars crew. They are considered low in self-disclosure ability, although this may be a result of being in maledominated occupations. Dindia and Allen (1992) have a comprehensive study showing that women are marginally more self-disclosing than men. Perhaps journaling is not for the average Mars crew, nor group therapy. However, even crew in these occupations can come to realize that, as Tamir and Mitchell (2012) claim, disclosing information about the self is intrinsically rewarding. Colleen McLaughlin

Barlow, artist-in-residence for a week at the Large Binocular Telescope in Arizona, had remarkable success in encouraging telescope engineers to express their creativity in art, as witnessed by one of the authors. The special environment of Mars may bring out unexpected talents and insight, surprising everyone. Pioneers of every era have experienced this.

Digitized, keyed, psychology workbooks on specific problem issues can be easily available for crew members who work their way through questions and examples at their own pace. The advantage is that no verbal response or writing is necessary, and yet the issues are covered by material written on Earth by expert therapists. In this way, crew members can become more self-aware and self-reflective concerning their mental status and social interactions, and they can do so without self-disclosure or journaling.

The Issue of Privacy. An important issue that can induce reluctance for self-disclosure is privacy. Recent concerns embodied in the European Union's 2018 General Data Protection Regulation have emphasized that privacy is a right for all humans, one to be lessened or given up only for the sake of some important goal. While in training, the crew may have become accustomed to Implantable Medical Devices that transmit data on their physiological condition, but that information does not include their most private thoughts. Analysis of stored physiological data will be a reflection, in part, of psychological status, like data used in lie detection. The crew will also have experienced lengthy interviews during training, so they are aware that psychological issues are important. They will know, too, that the balance between safety and security is not new (Halperin et al. 2008; Camara et al. 2015).

Still, the intrusiveness of medical implants and any type of psychological probing may be all the more acute for the crew in their close confinement on the trip to Mars and on the planet. "Brainjacking," the unauthorized manipulation of implants in the brain is even more horrific to contemplate (Pycroft et al. 2016), but some crew may fantasize about this if under great stress and if the symptoms of paranoia begin. Given these possibilities, every measure to maintain the trust of the crew should be taken. We will consider robotic modalities in the next section. Can these, which will generally be connected to a central computer system, also maintain sufficient privacy to keep the confidence of the crew? Without that confidence, it is conceivable that crew on Mars could "go rogue" and refuse help they need from mission support on Earth.

Robotics: Service Delivery without Verbal Feedback. Some people are prone to loneliness and depression more than others. Loneliness coupled with stressors becomes very debilitating according to Masi and colleagues (2011). These authors have identified and assessed four primary intervention strategies for treating loneliness: (1) improving social skills, (2) enhancing social support, (3) increasing opportunities for social contact, and (4) addressing maladaptive social cognition. It has been found that social robots, like a robotic dog, are as effective as a real dog in reducing loneliness (Banks et al. 2008).

Animal-assisted therapy with social robots can currently provide examples that may be transferable to Mars crew. Aaron Horowitz's robot duck, based on the duck that advertises for Aflac, is charming and effective (Holland 2018). Children communicate with social robots on a peer-to-peer basis and use them to express their feelings more freely. The article recalls that Jerry the Bear, an interactive companion for children with diabetes, was a predecessor to the duck. Experimentation by Japanese researchers started at AIST back in 1993 with Paro, a baby seal, again very cute, with big appealing eyes and soft fur, which was effective in reducing stress (Birks et al. 2016). Less expensive than these is Hasbro's Cat, now used with seniors (Newman 2016).

Some writers have raised ethical concerns about the use of social robots, especially for children, since they are unsophisticated, nonmoral agents. However, Calo and collaborators (2011) point out that the intention of the creators and providers points to the important moral issue. Similar questions have arisen over parents providing a teddy bear to their children, and this dilemma was solved long ago. We have all enjoyed the comfort of cuddly toys when young.

Adult, inpatient substance abusers also enjoy stuffed animals, and take them everywhere, including community meetings. In the absence of human social interaction for the young, the old, and the substance abuser, moral issues become less pressing. It is an idea worth considering to provide adult Mars crews with furry (or other) robodogs who can provide good company and a personal confidant.

AI Technologies: Services with Verbal Feedback. A significant upgrade in social robots is ones that converse, i.e., Fully Automated Conversational Agents or "chatbots." In a sense, the chatbot can substitute for a therapist, especially if there is sophisticated AI software that guides them. Kathleen Kara Fitzpatrick, Alison Darcy, and Molly Vierhile studied the effect of conversations with Woebot, a cartoon figure, by a group of some 70 college students, mostly women. While their sample was lopsided, "those in the Woebot group significantly reduced their symptoms of depression over the study period..." (2017, Fig. 2). An earlier study of "Ellie," an avatar-like female on a screen, showed something very interesting: People were more willing to reveal themselves when they were told Ellie was totally a robot than when they were told, falsely, that it was just an electronic puppet for a human therapist (Lucas et al. 2014).

An extension of the Ellie study (Pickard et al. 2016) showed that the more sensitive the information they revealed, the more comfortable the people were with Virtual Humans (VHs). This may be the solution to those reluctant to self-disclose, who may predominate in the Mars crew.

Therefore, chatbots and VHs, even in their early stages of development and testing, do work. When 3-D hologram versions of these counselors are developed, they will be portable and ready for Mars.

Robotic Vehicles, and Human Needs to Explore and Play. There are now robotic vehicles that already have an AI to control their movements safely in traffic. Soon, developers will have an AI that provides feedback through the "driving experience" of acceleration, cornering, and adventure, all the things that gave us a thrill in our first car. This type of added dimension would provide a sense of control in an all-too-robotic world. They may help increase the important feeling of wonder that will make life on Mars tolerable and hopefully pleasurable. Humans play, and technologies that encourage play with the goal of maintaining self-esteem and mental health, will be an important addition to the service delivery options.

Problem-Solving at the Cultural Level. Some of the group and community modalities we have mentioned to this point will allow the crew at a Mars hardship post to (1) notice how culture is changing, (2) decide whether the change is desirable, and (3) consider how to further implement a change. One advantage is that on Mars, if cultural problems are noticed, they will almost surely be connected in some way to one of the exogenous stressors we have previously listed—gravity, atmosphere, radiation, light, or dust—because they have changed the most. One of the purposes of culture is to provide readymade solutions to everyday problems, so humans do not need to think through routine problems each time they are encountered. Solutions become second nature, except perhaps on Mars. We address one particularly thorny issue in the following section on Spiritual Services Modalities—death and burial.

As we have emphasized, each crew member will come with a full cultural complement that will immediately begin changing when they interact with crew who have different cultures. Cultural differences may cause conflict, but the process of cultural synthesis cannot be stopped (although it can be guided). At times, cultural beliefs will be so covert and so deeply held that individual crew will not realize why they are reacting in certain ways. They will simply feel that "something is not right." These sentiments are understandable because culture defines for humans what is right and what is wrong, and it defines them differently for various groups. These understandings are passed on so early and so covertly, that many of us are not aware we hold them.

It will be useful to have a context in which to air cultural differences. On Earth, this is often at mealtime, partly because foods, tastes, and smells often call forth our most deeply held beliefs and oldest memories. Feeding remains one of the most intimate group contexts humans have, but it is also useful to remember that conflicts can result suddenly at meals, too, in part because of these old memories.

To date, government agencies of industrialized nations and international organizations have been notably unsuccessful in changing cultures of people for humanitarian purposes. (Two exceptions may be contraceptives and oral rehydration for infants in developing countries.) Two problems predominate: First, staff are sometimes convinced a native culture must be protected at all cost and they do not want to change it, and second, the methods for introducing new culture rarely work. There have been efforts to "technologize" the transfer of cultural values and practices, and various efforts at "technology transfer." Unlike psychological interventions, which target individuals and perhaps families, cultural interventions target entire groups, and changing a group's culture can be very, very complex. Nevertheless, survival on Mars will depend on the culture of the crew adapting to new and very difficult conditions, so it is worth considering what kinds of contexts and modalities might serve to, at minimum, review the issue.

The only context where culture has proven amenable to change is in businesses and large organizations with a specific, isolated cultural problem like ethnic discrimination or sexual harassment. For example, Katzenbak et al. (2012) give examples of the problems in changing an entrenched culture. They start with what is already working and outline five strategies for change: match strategy to culture; focus on a few critical shifts in behavior; honor the strengths of the existing culture; integrate formal and informal interventions; and, measure and monitor cultural evolution. To this, we would add, identify dysfunctional practices of an existing culture.

Changing culture is a difficult undertaking, even with the powerful motivation of profit. An intervention for a localized or specific conflict within a business can be more successful, since the players and factors are more limited, but still, the only solution may be letting employees go. Varnum and Grossmann (2017) describe "ecological pressures" that may cause a culture to change, and these pressures will be extreme for the crew on Mars. When life is on the line, crew will have great motivation to change even the most deeply held cultural beliefs. Program managers on Earth should be ready for unexpected cultural changes by human crews and communities on Mars.

Cross-Cultural Modes of Participation and Tension Release. To this point, we have mentioned robotic vehicles, friendly robodogs, an "art area", an anonymous electronic bulletin board for writing on stressors, and other modalities involving play and creativity that could transform into problem-solving activities. From our knowledge of events and organizations that cross-national and regional boundaries, we note that some types of group expression appear to somehow pass a "cultural boundary" test and are easily open to all people. Art is an obvious example, but so are games, music, dance, and sports. These are usually seen within a context of "recreation," and we are drawn to the origin of that word—to re-create—when we suggest provisions to facilitate recreational activities among Mars crew. The payoffs may well be reduced tension, better rest, and enhanced creativity in the thinking of the crew.

Teams for soccer and touch football conjure up versions of these sports with humans who skip. The humor resulting from these efforts may be as therapeutic as anything else. Similarly, dancing with a skip could also be a challenge. Or, a simple foot- (boot-) race. Games could dramatize daily efforts and accomplishments. New words would be devised, new goals could be set forth, and new payoffs would emerge as rewards for passing, or winning, or simply completing a task. As on Earth, teams are chosen, team names capsulize valued qualities of the members, and success is simply defined. These types of recreational activities are requisite by human communities. When Martian crew begin devising their own games, it will be clear that a Martian "community" is beginning to form.

2.5 Spiritual Services Modalities

We have written extensively on the evolution of moral and religious capacities, so our comments on spiritual services on Mars will reflect what we understand about the requirements for those higher, human cognitive capacities. Both are strongly social and dependent on group activities that are physically and temporally set apart from normal interaction. This is not to say that humans cannot reason morally and reflect on religious themes by themselves. They do, and solitary contemplation is important. However, moral and religious capacities arose as mechanisms to support human social groups characterized by culture that is shared. We estimate that the biological capacity for culture began to emerge around 8–10 million years ago [Note 1], but the full emergence of religious capacity is much more recent, around 150,000 years ago.

Religious rituals are practiced by humans most often in groups and rely on a "faith community" to arbitrate important decisions about right and wrong, appropriate and inappropriate, and real versus unreal issues. Given these parameters, we make the following observations on needed "spiritual services" on Mars. We focus on two main topics: (1) sacred space; and (2) an example of human funeral and burial rites. Like other rites of passage, they tend to become culturally standardized over time in isolated communities. We choose this particular rite of passage because it very likely will be the first type of rite of passage for humans on Mars—unless, of course, someone decides to get married! It is always a possibility, but, as among the first Peace Corps workers, discouraged. When humans are so far away from "home," their perspective on mate selection is understandably skewed. Eventually, Mars will be "home" and marriages will take place. The big question is the advisability of human reproduction, given the exogenous stressors listed at the start of our chapter.

Sacred Space. We recommend that a "sacred space" be set aside or converted on both the trip out to Mars and in the first habitations. The space can be a commonly used space like a mess hall or cafeteria, which receives a special, removable décor when crew are assembled for "services" or reflections about spiritual topics. Indeed, that décor could be so modest in size and so portable, that a crewmate could borrow it to pray or reflect, and then return it for group use. The assemblies for religious purposes should never be required. Some humans do not, whether by choice or inclination, participate in religious activities. This should be anticipated, since proportions of atheists and agnostics cannot be entirely foreseen. And, decisions about belief may change among individual crew, in one direction or the other.

On Mars, sacred spaces will be naturally sought out for their beauty or their ability to call forth awe that usually attends religious belief. Planners should expect these locations to be identified, and not be surprised when one, then a few more, of the crew decide to seek solace in sacred surroundings. Humans have been doing this for hundreds of thousands of years.

Crew should be warned about religious conflicts and encouraged to seek commonality in prayer or simple reflection. Readings, music, and singing need not always be specific to a chosen religion, just as they need not be specific to a certain culture or nationality.

An Example of Accommodation to Mars: Death and Burial Rituals. We will round out our brief analysis of spiritual services by giving an example that stems directly from several of the exogenous factors mentioned at the beginning. When a human dies, the group is always affected, and consequently the passing is marked with ritual that reflects the belief system of the group. Since the earliest Martians will still maintain beliefs born on Earth, there will be a hodgepodge of suggestions for how to handle a death. Some of these relate to sanitation; some relate to grief; and some relate to their new home, Mars. Some of the specific aspects discussed here are adapted from a fiction story written by one of the authors (Rappaport 2002).

The story is set in a time when the first Martian crews have already transitioned into "communities," so there has been time to organize and establish rules for handling the dead.

We first emphasize sanitation. A dead body will immediately begin to decompose inside habitations that are sufficiently warm for humans, so immediate disposal of the body is required. In the story we use for some of these details, a female body is handled by the females, and a male body is handled by males.

Next, we emphasize a usefulness and therefore, a need for the body. Mars has no natural organic material. It has no soil in which to grow plants. It has no substrate on which to grow fungi. All possible organic material is recycled through a waste management system, and then, it is composted. For the community on which the story centers, there has been sufficient time to decide what to do with human remains. Most of the body is recycled in the compost system, leaving just one hand to bury in a community burial plot. In this way, a special organ of the human is buried in a plot, and the rest is reused to help support the community.

Finally, we emphasize ritual in a human group that has already combined aspects of different faiths. The ritual is very simple. The "captain" or "governor" of the community takes the hand of the deceased and buries it deeply in the regolith, with a marker. These have been prepared ahead of time because of the very high-speed winds that blow on Mars. The hand of the deceased is buried securely so that it will not be uncovered, and the marker is securely fixed.

After burial of the hand by the leader, the entire community "suits up" in the cafeteria, forms a long line, and one by one exits the habitation, walks by the burial, and then slowly reenters the habitation. There is no sermon, no songs, no expressed beliefs, except at the meal that follows when the deceased is fondly remembered by the assembled community.

A human burial is rarely more simple than that, but still, its formality signals an important change in the community. A person is gone. He or she is remembered. He or she has left part of the body, the hand, to mark the passage. The community grieves and returns to normal. Higher Primates all grieve, to various extents. It is to be expected, planned for, and ritual should occur that can focus that grief. It will allow the living to go on normally.

In this simple scenario, we see the importance of spiritual reflection, stopping to mark the passage of a community member, and taking time to remember. Martian funeral practices will evolve differently perhaps, but some of these essential elements will be present.

It is this type of cultural creativity that we will eventually observe among Mars crews and early communities. One of the most important functions of rituals, within a spiritual context, is to provide hope and faith in the future for the crew or community living in very challenging conditions.

2.6 Note

Note 1. We use the horizon of 8–10 million years ago because that is the time of the split of humans from the line to the chimpanzees and bonobos. The latter evidence culture weakly and humans show it strongly. However, the biological basis for culture—at least in its most rudimentary form—should date to this early time. Recall that sociality had emerged among primates 55–65 mya, and had been running along nicely for a long time. Culture took a different biology, and it likely emerged in some isolated groups of late Miocene apes.

References

Acevedo, B. P., Aron, E. N., Aron, A., Sangster, M.-D., Collins, N., & Brown, L. L. (2014). The highly sensitive brain: An fMRI study of sensory processing sensitivity and response to others' emotions. *Brain and Behavior, 4,* 580–594.

Ackermann, M., & van den Bogert, A. J. (2012). Predictive simulation of gait at low gravity reveals skipping as the preferred locomotion strategy. *Journal of Biomechanics, 45*(7), 1293–1298.

Aron, E. N., Aron, A., & Jagiellowicz, J. (2012). Sensory processing sensitivity: A review in the light of the evolution of biological responsibility. *Personality and Social Psychology Review, 16,* 262–282.

Banks, M. R., Willoughby, L. M., & Banks, W. A. (2008). Animal-assisted therapy and loneliness in nursing homes: Use of robotic versus living dogs. *Journal of American Medical Directors Association, 9,* 173–177.

Birks, M., Bodak, M., Barlas, J., Harwood, J., & Pether, M. (2016). Robotic seals as therapeutic tools in an aged care facility: A qualitative study. *Journal of Aging Research*, Article ID 8569602, 7 pages.

Braddock, M. (2018). Concepts for deep space travel: From warp drives and hibernation to world ships and cryogenics. *Current Trends Biomedical Engineering & Bioscience, 12*(5), 555847.

Bradford, J., Schaffer, M., & Talk, D. (2014). *Torpor inducing transfer habitat for human stasis to mars*. Final report prepared for the National Aeronautics and Space Administration (NASA), Grant Number NNX13AP82G. May 11.

Calo, C. J., Hunt-Bull, N., Lewis, L., & Metzler, T. (2011). Ethical implications of using the Paro robot with a focus on dementia patient care. In Proceedings of the AAAI Workshop on Human-Robot Interaction in Elder Care, WS-2011-2012, San Francisco, California, USA (pp. 20–24).

Camara, C., Peris-Lopez, P., & Tapiador, J. E. (2015). Security and privacy issues in implantable medical devices: A comprehensive survey. *Journal of Biomedical Informatics, 55,* 272–289.

Cochran, G., & Harpending, H. (2010). *The 10,000 year explosion: How civilization accelerated human evolution*. New York: Basic Books.

Cockell, C. (2002). Mars is an awful place to live. *Interdisciplinary Science Reviews, 27*(1), 32–38.

Cockell, C. S. (Ed.). (2016). *Dissent, revolution and liberty beyond earth*. New York: Springer Publishing.

Dindia, K., & Allen, M. (1992). Sex differences in self-disclosure: A meta-analysis. *Psychological Bulletin, 112,* 106–124.

Dutton, D. (2009). *The art instinct: Beauty, pleasure, and human evolution*. New York: Bloomsbury.

Forsyth, D. R. (2015). Group psychotherapy, clinical psychology. In J. D. Wright (Ed.), International encyclopedia of the social & behavioral sciences (2nd ed., Vol. 10, pp. 428–433). Oxford, UK: Elsevier.

Gaudin, S. (2014). When astronauts breathe on mars, they'll thank MIT professor; Researchers building device that will create oxygen on the red planet. Computerworld (Online. August 7, 2018).

Halperin, D., Heydt-Benjamin, T. S., Kevin, F., Kohno, T., & Maisel, W. H. (2008). Security and privacy for implantable medical devices. *IEEE Pervasive Computing, 7,* 30–39.

Hawks, J., Wang, E. T., Cochran, G. M., Harpending, H. C., & Moyzis, R. K. (2007). Recent acceleration of human adaptive evolution. *Proceedings of the National Academy of Sciences, 104*(52), 20753–20758.

Holland, P. (2018). 'My Special Aflac Duck' is a robot to help kids with cancer. Retrieved June 19, 2018, from https://www.cnet.com/news/my-special-aflac-duck-robot-helps-kids-through-cancer/.

Hublin, J.-J., Ben-Ncer, A., Bailey, S. E., Freidline, S. E., Neubauer, S., Skinner, M. M., et al. (2017). New fossils from Jebel Irhoud, Morocco and the Pan-African origin of homo sapiens. *Nature, 546,* 289–292.

Impey, C. (2018). Personal communication.

Kanas, N. (2014). The new martians. A scientific novel. Switzerland: Springer International Publishing (Digital edition).

Kanas, N. (2015). *Humans in space; The psychological hurdles*. Chichester, U.K.: Springer Praxis Books.

Kara, F. K., Darcy, A., & Vierhile, M. (2017). Delivering cognitive behavior therapy to young adults with symptoms of depression and anxiety using a fully automated conversational agent (Woebot): A randomized controlled trial. *JMIR Mental Health, 4*(2), e19.

Katzenbach, J. R., Steffen, I., & Kronley, C. (2012). Cultural change that sticks. Harvard Business Review, Vol. 90, pp. 110–117.

Kazdin, A. E., & Blasé, S. L. (2011). Rebooting psychotherapy research and practice to reduce the burden of mental illness. *Perspectives on Psychological Science, 6*(1), 21–37.

Kino, E. F. (1919). Kino's historical memoir of Pimeria Alta. (H. E. Bolton, Trans.). Bedford, MA: Applewood Books. (Original: 1683–1711).

Lucas, G. M., Gratch, J., King, A., & Morency, L.-P. (2014). It's only a computer: Virtual humans increase willingness to disclose. *Computers in Human Behavior, 37,* 94–100.

Masi, C. M., Chen, H.-Y., Hawkley, L. C., & Cacioppo, J. T. (2011). A meta-analysis of interventions to reduce loneliness. *Personality and Social Psychology Review, 15*(3), 219–266.

NASA, National Aeronautics and Space Administration. (2011). Psychology of space exploration: Contemporary research in historical perspective. In: D. A. Vakoch (Ed.), *The NASA History Series, NASA SP-2011-4411*, Washington, D.C.: USGPO.

NASA, National Aeronautics and Space Administration. (2018). NASA twins study confirms preliminary findings. Reported online by M. Edwards & L. Abadie, updated March 5, 2018 (by T. Gushanas (Ed.).

Newman, A. (2016). Therapy cats for dementia patients, batteries included. https://www.nytimes.com/2016/12/15/nyregion/robotic-therapy-cats-dementia.html; Oviedo, L. (2019). Raising the right questions. In this volume.

Pickard, M. D., Roster, C. A., & Chen, Y. (2016). Revealing sensitive information in personal interviews: Is self-disclosure easier with humans or avatars and under what conditions? *Computers in Human Behavior, 65,* 23–30.

Pycroft, L., Boccard, S. G., Owen, Sarah L. F., Stein, J. F., Fitzgerald, J. J., Green, A. L., et al. (2016). Brainjacking: Implant security issues in invasive neuromodulation. *World Neurosurgery, 92,* 454–462.

Rappaport, M. B. (2002). Hank chin; Scientist, seer, rebel, and founder of the first martian republic. Book II of The Nanotech Trilogy. Manuscript.

Schaffer, M., Bradford, J., & Talk, D. (2016). A feasible, near-term approach to human stasis for long-duration deep space missions. In 67th International Astronautical Congress (IAC), Guadalajara, Mexico (pp. 1–17).

Seto, K. C., Golden, J. S., Alberti, M., & Turner II, B. L. (2017). Sustainability in an urbanizing planet. Proceedings of the National Academy of Sciences, 114(34), 8935–8938.
Snyder, M. (2013). Journaling. In R. Lindquist, M. Snyder, & M. F. Tracy (Eds.), Complementary & alternative therapies in nursing (7th ed., pp. 205–215). Chichester, U.K.: Springer.
Solomon, S. (2016). *Future humans; inside the science of our continuing evolution*. New Haven: Yale University Press.
Szocik, K. (2016). Religion in a future mars colony? Spaceflight. November.
Szocik, K., Marques, R. E., Abood, S., LysenkoRyba, A. K. K., & Minich, D. (2018). Biological and social challenges of human reproduction in a long-term mars base. *Futures, 100,* 56–62.
Szocik, K., Campa, R., Rappaport, M. B., & Corbally, C. (In press). The special case of modifications to counter bone loss for manned mars missions.
Tamir, D. I., & Mitchell, J. P. (2012). Disclosing information about the self is intrinsically rewarding. *Proceedings of the National Academy of Sciences, 109,* 8038–8043.
Todd, R. M., Ehlers, M. R., Müller, D. J., Robertson, A., Palombo, D. J., Freeman, N., Levine, B., Anderson, A. K. (2015). Neurogenetic variations in norepinephrine availability enhance perceptual vividness. Journal of Neuroscience, 35, 6506–6516.
Ullrich, P., & Lutgendorf, S. K. (2002). Journaling about stressful events: Effects of cognitive processing and emotional expression. *Annals of Behavioral Medicine, 24,* 244–250.
Varnum, Michael E. W., & Grossmann, I. (2017). Cultural change: The how and the why. *Perspectives on Psychological Science, 12,* 956–972.

Chapter 3
Enabling and Enhancing Human Health and Performance for Mars Colonies: Smart Spacecraft and Smart Habitats

Mark Shelhamer

Abstract Crews on extended space flights—such as missions to Mars for exploration and settlement—will be faced with unprecedented challenges. In addition to recognized concerns about human health and performance that are already addressed through various mitigation strategies, there will be problems that have not yet been identified, some of which will be a consequence of the web of interactions among the many subsystems that make up such a complex undertaking. The people on these missions must also, due to remoteness and isolation from Earth, have the tools to assess and correct these problems autonomously. An approach to this situation is described here, which formulates the myriad subsystems and their interactions as a network (drawing on concepts from complexity theory). Each subsystem is a node, and the interconnections are links. Mathematical tools of complexity and network theory can be applied to assess the ability of the network to accomplish its given tasks (performance monitoring) and to maintain performance in the face of failures or perturbations (resilience).

3.1 Introduction

We could send people to Mars right now. Except for the slight problem of having sufficiently powerful rockets, we know enough to send people to Mars and possibly even have them survive for a long time. The technical challenges are great but not insurmountable. Of course, there are political issues involved, which are perhaps even more challenging, as pointed out long ago by Wernher von Braun and others.

There is, however, more to it than rockets and politics. People sent to Mars now or in the foreseeable future might *survive*, but they would not *thrive*. They might be able to live long enough to later return to Earth or perhaps to stay and reproduce. But they would not be able to do the very things for which they would be sent: explore, build infrastructure, conduct meaningful science, settle and colonize, and make the experience easier for future generations. The difficulties of a mission to Mars are

M. Shelhamer (✉)
Johns Hopkins University, Baltimore, MD, UK
e-mail: mshelhamer@jhu.edu

K. Szocik (ed.), *The Human Factor in a Mission to Mars*,
Space and Society, https://doi.org/10.1007/978-3-030-02059-0_3

numerous and have been enumerated many times by many people. The health hazards alone are legion: increased risk of cancer, cardiovascular inflammation, and acute cognitive deficits, all due to radiation; sensorimotor ataxia due to altered gravity; possible decreases in bone, muscle, and cardiac capability from those and other causes. There will also be issues of psychological health: detachment from Earth, problems with mood and motivation, interpersonal challenges stemming from small groups in confined spaces for long periods of time. These might be exacerbated by poor design of the spacecraft and habitat, improper training and allocation of key tasks, and lack of proper planning in the overall design of the mission. This complexity of interacting components can lead to failure modes and breakdowns in mission resilience that are difficult to predict, and for which it is difficult to plan mitigation strategies in advance. It is such unknown unknowns—deriving from complex interactions—that will be the biggest problem on such journeys.

So, since we as mission planners or as a society cannot think of everything in advance, we must provide these explorers and settlers with the tools they need to deal with the unexpected, some of which will arise from unpredicted (but maybe not unpredictable) confluences of factors, each of which individually might have been adequately mitigated. They will need tools for *resilience*—tangible tools and also conceptual tools—and it is the latter that we will discuss here (Shelhamer 2015).

3.2 Spaceflight is Dangerous

Consider an actual case of multiple-system breakdown that led to a near-disaster in space (Shayler 2000). In 1997, a Progress resupply ship, on routine approach to provide material to a crew on the Mir space station, crashed into Mir. This is about as serious an issue as can occur in orbital space flight, and almost led to evacuation of the station. Let us briefly review some of the circumstances that preceded this specific event.

Earlier in the mission, there had been a fire on Mir (a near-disaster of its own), for which mission control placed initial blame one of the on-board crew members. There was also an attempt to undock a resupply ship that was already attached to Mir, and re-dock it at another port, partly to free up a desired port but also as a means for the crew to practice the docking operations. These docking operations are performed by the crew commander with on-board controls and a video display from a camera in the resupply ship. This practice re-docking was a failure.

At various other times in that mission, an oxygen separator failed, as did a CO_2 scrubber and its backup system; there were problems with availability and reliability of spare parts for the separator. For a time, a spacecraft attitude sensor failed, placing the station in a gravity-gradient mode for attitude control, leading to a reduction in available on-board power due to nonoptimal orientation of the solar panels with respect to the sun. This also contributed to an increase in internal temperature, limiting the ability of the crew to exercise (which has benefits for physical as well as psychological health). There had been a leak of coolant into the station, producing an allergic reaction when it got into the eye of one of the crew. There had been a toilet failure, the passageways were cluttered with equipment, and various atmospheric gas fittings were incompatible with each other. As might be expected from all this, the crew was experiencing a very high workload, and falling behind in scheduled tasks; they had had no free weekend in three months.

American astronauts were taking turns as crew members on Mir at that time, but the interaction between Russian cosmonauts and American astronauts was not yet as coordinated as it would be in the ISS era. There was in fact some lack of crew cohesion in that the Russian cosmonauts (at least at first) shared little with the American astronauts, as Mir was a Russian space station of which they were justifiably proud.

As for the Progress docking itself, crew members helping the commander had restricted views of the approaching resupply ship through Mir's small observation windows. The commander had a video image provided by a camera on Progress, but unclear instructions on what to expect from the docking initiation. The video imagery had poor visual reference points and poor lighting, and Progress itself had poor rocket thruster performance. Finally, the commander had not practiced a docking for 130 days.

It would be easy to attribute the ensuing crash to any one of a number of causes. Crew fatigue and health were clearly critical issues, perhaps the most significant. But spaceflight crews work through fatigue all the time. Interactions of the crew members with each other and with ground control were contributing factors, as were various human factor issues regarding manual control of the approaching Progress spacecraft. Based on the earlier failed docking attempt and the high competence level expected of flight crews, performance anxiety must have been extremely high. It is tempting to try to identify the single most important of these causes, as in a traditional root cause analysis. The true fault, however, lies in the fact that these various factors interact with each other, and several were breaking down simultaneously. "It was not the fault of one person or element, but a combination of several actions of a variety of people and by different hardware and software" (Shayler 2000). Not only that, but perhaps, with the correct monitoring and tracking, this situation might have been predicted and avoided through proper interventions.

3.3 Lessons from Disaster

What does this episode tell us? First, despite extensive training and high motivation, bad things can combine and build up to intolerable levels on something as demanding as a space flight (or colony or settlement). Second, it is the pernicious combination of several nominally disparate factors that can present the biggest problem. If this is true for something as relatively simple as a Mir mission in low Earth orbit, imagine the potential for disaster on something as complex as a Mars mission, especially one that involves extended stays and colonization. However, this example presents a lesson in that such critical breakdowns might be anticipated and prevented. How might we do such a thing?

Let us now imagine an alternative scenario. Assume that late in a planetary exploration mission, one crew member is showing subtle but slowed performance on a routine maintenance task, as determined by an instrumented workspace (with sensors that track body motion and integrate it with task difficulty and historical performance levels for this person on this task). Other sensors indicate that this same person is having difficulty relating to the commander, which is unusual; this is detected with voice stress analysis and relative position/orientation tracking of the individuals involved. Unobtrusive and noninvasive assessment of key physiological parameters detects the initial indications of a fever and electrolyte imbalance in this person. In response to these observations, schedule changes are made to provide the person with more individual tasks (a form of mild quarantine), local CO_2 level is decreased, and supplements and medication are added to his or her food. This is all done automatically with no crew action. If more extreme interventions were to become necessary (e.g., high medication doses), the medical officer, commander, and affected crewmember could be notified and asked for approval. Critically, the effects of these changes over the next hours or days are then tracked by the same sensor suite, and the interventions are tapered off or altered as needed based on this feedback. Potential disaster is averted since interventions were made at the early stages of an impending multifactor problem.

3.4 A Solution

What do these two example systems have in common with possible future Mars colonies, or human space flight more generally? They all depend on the proper functional integration of multiple disparate factors for success. These factors span a range from human physiology, psychology, and performance, to spacecraft and mission design, task allocation and training procedures, and many more. They have been captured in a Contributing Factor Map (CFM: Mindock and Klaus 2014), which provides a systematic taxonomy of these factors across the many domains (Fig. 3.1).

The key to proper functional integration of these many factors in space flight is that the various subsystems (contributing factors) must work together. They must be integrated. There is some nuance to this. It does not mean that everything is very strongly connected to everything else; that would lead to a cascade of failures (domino effect). It also does not mean that the subsystems are sparsely connected; then there would be little to no redundancy, no alternative solutions when subsystems failures occurred, and hence no resilience. There must be the proper arrangement of interconnections between factors, and these connections must be flexible. There are times, for example, when an astronaut will welcome a great deal of automated assistance from robots and computers, as for example when workload is high and the tasks to be carried out are well-defined. When novel and potentially dangerous

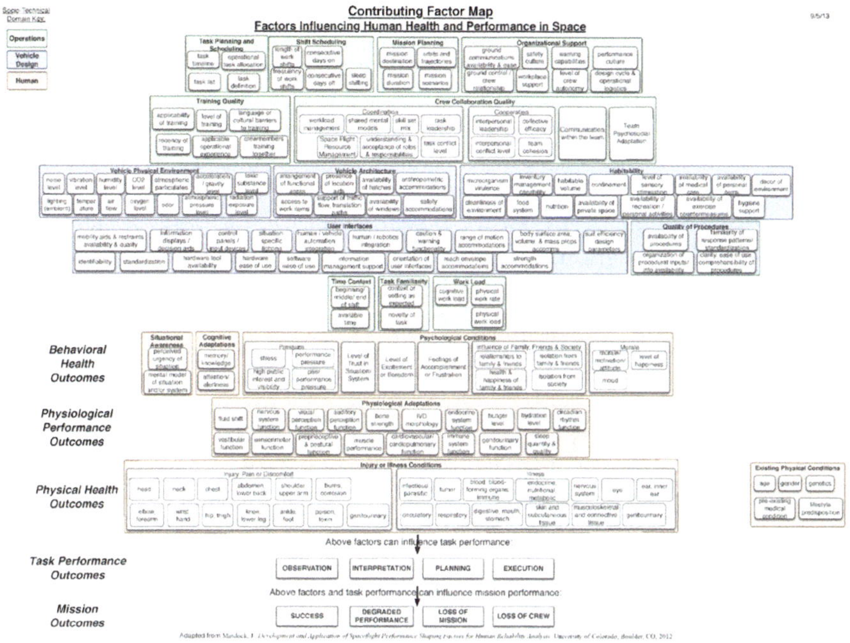

Fig. 3.1 Contributing Factor Map (CFM), a taxonomy of the major factors that contribute to a successful human space flight. (Based on Mindock 2012, used with permission of the author.)

situations are encountered, on the other hand, and human decision-making is critical, then proper decision support, information flow, and situational awareness become the overriding factors. The automated systems should then exercise great care before they attempt to take autonomous actions that the human crew might not appreciate. Thus, the links between the various subsystems must be modulated in response to the needs of the current mission situation at any time.

Implementation of such a strategy is not a trivial undertaking. What is needed is a solid systematic understanding of interconnections between factors and how they lead to resilience or the loss of it. It is not sufficient just to say that this is connected to that, and so on. Some rigor is needed. In that way, the most important factors can be identified and measured, their interconnections characterized and tracked, and all of this information can feed a decision-making algorithm that estimates the current state of mission resilience or fragility. Some of the tools to do this are already in place but many have yet to be developed.

3.5 Some Specifics

A glance at the Contributing Factor Map (CFM) shows that many of the factors in question can be measured with the appropriate sensors. Some things, like the vehicle environment, can be monitored with sensors that are available now, and in fact the internal environments of all human spacecraft are monitored in this way. Similarly, a great many sensors already exist to monitor aspects of human physiology: heart rate, respiration, motion, metabolism, and many others. Development of new sensors is an active area of research and more sensors—both invasive and noninvasive—are becoming available all the time. It is even possible to monitor many things that might not appear to be amenable to this type of monitoring. For example, some aspects of inter-personal interaction and team function can be assessed with objective measures such as patterns of voice communication (turn-taking, etc.), and sociometric badges can track individuals as they interact, showing when people are forming cliques, if someone is becoming an outlier or loner, how often people congregate, and similar patterns that can represent healthy or unhealthy group behavior. Some of the factors in the CFM, however, are not so easily assessed, and new sensors, means of assessment, or surrogate measures may have to be developed.

This then brings us to the hard part: what are the interconnections between factors and how can they be tracked? This goes beyond simply tracking each parameter individually (which can be challenging enough). Some problems might indeed be detectable by flagging single parameters (factors) that are outside of their normal range (whatever has been determined to be normal for that person or factor at that phase of the mission). The real power of the proposed formulation, however, relies on capturing, characterizing, and tracking the interactions and interconnections between parameters.

Some of these cross-factor relationships and interactions are straightforward, at least in principle. Consider that an increase in cabin CO_2 level should have well-established effects on the crew: lethargy, headache, irritability, cognitive slowing, and others. The spectrum and severity of these effects will differ between crew members, and this information can be captured through simulations and other baseline measures. By monitoring CO_2 and associated patterns of crew activity during a space flight (including such related things as medication usage), the links between these factors can be further established and refined. At some point in the mission, if CO_2 becomes elevated and the expected crew effects do *not* occur, then the interactions have been altered. Whether or not this is of concern is the larger and more important issue, as we discuss below.

Establishing links (interactions) between factors enables a giant leap: it turns the relatively simple *taxonomy* of the CFM into a complex *network*. The boxes in the CFM (major factors) are nodes in the network, and the interactions are the links between these nodes. This allows us to take advantage of the large and growing body of research into networks, complex systems, and complex networks. In particular, it holds out promise that *resilience* might be mathematically defined and tracked during the course of a mission (Gao et al. 2016), as parameters and their interactions change.

At this point, perhaps it is helpful to discuss what is meant by resilience. In our context, we mean the ability to recover from perturbations and maintain adequate performance—the ability to accomplish mission goals when things go wrong. This is a consequence of a delicate balance of redundancy and isolation. If subsystems (factors in the CFM or nodes in the network) are too highly interconnected, then a failure in one can lead to a cascade of failures in other subsystems. If subsystems are too sparsely interconnected (they are isolated from each other), then there are insufficient alternatives and backups when something goes wrong. The right balance should be achievable through careful design of the mission and spacecraft, training of the crew, and attention to concepts of resilience engineering (Hollnagel et al. 2007) such as described here. This form of resilience might even be defined mathematically (Gao et al. 2016) and monitored for changes during the mission. In that case, changes in the network—the overall system-of-systems of multiple interacting factors that make up the mission—might be detected with the proper sensors and algorithms (mathematical models). If these changes decrease resilience, as defined by the mathematical formalism, then interventions can be suggested to bring resilience back. This is the ultimate goal. The model can also run "what-if" scenarios on itself. With each substantial change in parameters or interactions, multiple simulated scenarios can be run on the model to assess the resilience of the current model state to further perturbations.

Finally, there is the possibility of using the results of the algorithm to make changes to the spacecraft or habitat in the course of the mission, in order to maintain and enhance resilience and performance. An example was provided previously. Work is needed in this area to determine the types of reconfigurations that might be accomplished, what might be done automatically and what might be done by the crew with automation assistance, and what might only be performed by the crew after advice or suggestion from the computer model. It might simply be sufficient to provide information to the crew (or just the commander or the medical officer) to indicate that a problem might be developing—a problem whose early indications may be too subtle for the human crew to detect. This is a critical point and a key feature of the entire endeavor: slight changes in interactions between factors might be an indication of a developing problem, which could be averted through early intervention.

3.6 Too Much?

Would crews accept this type of "assistance"? Those people who have the "right stuff" to venture into space are not the kinds of people who typically like to be watched all the time and told what to do. Is then this whole scenario too much like Big Brother? Will the crew, far from Earth and free to do as it pleases, defeat any attempts at monitoring, ignore any computer-provided suggestions, or even disable the system that has been designed to help them?

Hopefully not. There are two primary reasons to think this. First, the crew should be trained to properly understand the dangers that they face and the role of an on-board system to help them avoid these dangers. This system then will not be perceived as taking independence from the crew, but rather as granting them great autonomy—from far-away mission control, for example, which cannot have real-time knowledge of the situation on-board. The crew is free to ignore the automated advice and suggestions, ask for more information, or query the ground for further assistance if time permits. But at all times they have the ultimate authority to accept or ignore any suggested interventions—interventions that are designed to help the crew to help themselves. This aids their autonomy, which any space-worthy crew should welcome. Second, future flight crews will be comfortable with newer information technologies such as artificial intelligence, which would form one of the processing stages of the embedded system. AI is already making significant inroads in many areas on earth, and as this continues, future tech-savvy astronauts will be more accepting of it.

3.7 Conclusion

Many systems are resilient. The Earth's ecosystem is, on the whole, resilient, as are human societies. These, however, have evolved and developed over millennia or generations. The first Mars colonists will not have the luxury of such a timescale, and the logistics of resupply (of people and consumables) mean that the survival of these trans-planetary pioneers will depend almost solely on their own efforts: their resilience as individuals, as crews, and as an overall mission. They must be provided with the tools to self-organize in the face of the unexpected. Planning now for dealing with unknowable critical events can make it the case that "we should have seen this coming" can be turned into "we saw this coming…and we did something about it."

References

Gao, J., Barzel, B., & Barabási, A. L. (2016). Universal resilience patterns in complex networks. *Nature, 530,* 307–312.

Hollnagel, E., Woods, D. D., & Leveson, N. (Eds.). (2007). *Resilience engineering: Concepts and precepts*. Ashgate Publishing, Ltd.

Mindock, J. (2012). *Development and application of spaceflight performance shaping factors for human reliability analysis*. Aerospace Engineering Sciences Graduate Theses & Dissertations. University of Colorado, Boulder.

Mindock, J. A., & Klaus, D. M. (2014). Contributing factor map: A taxonomy of influences on human performance and health in space. *IEEE Transactions on Human-Machine Systems, 44,* 591–602.

Shayler, D. (2000). *Disasters and accidents in manned spaceflight*. Springer Science & Business Media.

Shelhamer, M. (2015). A call for research to assess and promote functional resilience in astronaut crews. *American Journal of Physiology, 120,* 471–472.

Chapter 4
The Impact of Space Weather on Human Missions to Mars: The Need for Good Engineering and Good Forecasts

Mike Hapgood

Abstract Space weather is a natural hazard that can adversely affect human missions to Mars, primarily through the adverse of impacts of energetic particles on spacecraft systems and on the health of astronauts. This chapter starts with an overview of these effects, and of their potential to disrupt human missions. This is followed by a discussion of the processes that generate these particles, highlighting that human missions to Mars must consider two distinct classes of particles: a slowly changing low-flux background of very energetic particles originating outside the solar system (galactic cosmic rays) and episodic bursts of high fluxes of slightly lower energy particles originating from the Sun (solar radiation storms). This distinction suggests that mission design should consider a mixed approach to mitigating the effects of space weather: (1) resilient design, including shielding to provide continuous mitigation of effects caused by the low-flux background; and (2) operational measures (sheltering, human troubleshooting) to mitigate the effects caused by radiation storms. The chapter also shows how our scientific understanding of space weather can guide efforts to mitigate these adverse effects: e.g. by flying to Mars when cosmic ray fluxes are lowest; by enabling space weather experts on Earth to provide targeted space weather forecasts so that astronauts are aware of when there is a significant risk that they will encounter a radiation storm; and by flying on-board particle instruments so that astronaut can control how they respond to the actual start and end of radiation storms. Finally, the chapter outlines the observing capabilities needed to support space weather forecasts targeted on human missions to Mars, showing that this will be synergistic with concepts now being developed for better forecasting of space weather at Earth.

M. Hapgood (✉)
RAL Space, STFC Rutherford Appleton Laboratory, Harwell Campus,
Didcot, Oxfordshire OX11 0QX, UK
e-mail: mike.hapgood@stfc.ac.uk

K. Szocik (ed.), *The Human Factor in a Mission to Mars*,
Space and Society, https://doi.org/10.1007/978-3-030-02059-0_4

4.1 An Introduction to Space Weather Effects

Space weather has gained much public attention over the past decade as we have begun to appreciate how it can disrupt many technological infrastructures that have become critical to the smooth running of everyday life here on Earth. Thus, governments around the world are now including space weather in the list of natural hazards for which they must prepare. In this chapter, we move the focus away from Earth to provide an overview on how space weather is a natural hazard that can adversely affect human missions to Mars, and outline how good engineering and good forecasts can mitigate those adverse effects.

4.1.1 Space Weather Effects on Satellites

As its name might suggest, space weather is an issue for which satellites and space missions must be prepared. Indeed modern satellites are designed and built to have high resilience to a range of space weather effects, in particular, exposure of satellites to energetic charged particles. In discussing these particular effects it is important to recognise that the space around and between the planets of our solar system is not empty. These regions of space are filled by a variety of tenuous plasmas, the fourth state of matter where neutral atoms have been broken down into negatively charged electrons and positively charged ions. These plasmas are very tenuous, typically only a few million particles in each cubic metre, more tenuous than what we would often regard as vacuum in a laboratory here on Earth. But the tenuous nature of these plasmas makes them behave in ways that are different to a gas. The particles in a gas interact via collisions (e.g. an oxygen molecule in the air at sea-level experiences billions of collisions per second). But particles in space plasmas rarely approach each other, and instead interact at long range (compared to particle size) via the Coulomb force. This leads to a wide range of collective behaviours that manifest, for example as waves, turbulence and shocks in the plasma, and in its associated electric and magnetic fields. But, most important for this chapter, the lack of collisions means that space plasmas are almost never in thermal equilibrium. Thus, space plasma particles do not exhibit a Maxwellian energy distribution, instead, their distributions exhibit long high-energy tails, often taking the form of a power law as shown in Fig. 4.1. The particles in these high-energy tails can travel long distances across the solar system without experiencing inelastic collisions that cause them to lose significant energy, though they may experience elastic collisions which change their direction. It is this property that allows solar energetic particles to travel out into interplanetary space where they are a major space weather risk, as we will see below, but also allows their angular distribution to become isotropic as particles are scattered by elastic collisions. This has profound implications for the impact of SEP events on the Martian environment as shown from measurements in, and modelling of, that environment (Luhmann et al. 2007). In the early phases of an event, the SEP

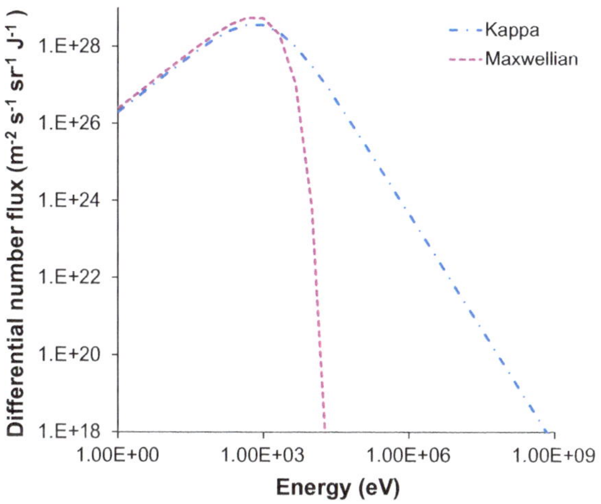

Fig. 4.1 Most space plasma environments are collisionless, so the velocity distributions of particles such as electrons and protons take the form of a Kappa distribution (Livadiotis and McComas 2013) as shown by the blue dot–dash curve. These distributions have long power law tails at higher energies, as well as high fluxes around the average particle energy. This example shows Kappa and Maxwellian distributions for an electron population with average energy ("temperature") of 1 keV and number density of 1×10^6 m^{-3}; and for the Kappa distribution, with a Kappa value of 2 (corresponding to a power law exponent of -2 in this figure)

distribution is anisotropic with the highest fluxes flowing away from the Sun along the interplanetary magnetic field. So, at this stage, the solid body of Mars can act a radiation shield reducing the SEP fluxes on the side of the planet opposite to the peak of the arriving fluxes. But this shielding effect becomes less significant as the event proceeds and the SEP fluxes become isotropic and envelope the whole of Mars.

Energetic charged particles have a range of impacts of satellites:

- Electrons with energies up to several tens of keV can deposit electrical charges on the surfaces of satellites. If there are significant fluxes of electrons at these energies, this charge can accumulate so that those surfaces acquire electrical potentials of thousands of volts. Different potentials on separate parts of these surfaces can lead to discharges that can damage satellite systems, and that could be a hazard to astronauts working outside a spacecraft. This charging is generally mitigated by ensuring, wherever possible, all the surfaces of a satellite are conductive and electrically grounded to the frame of the satellite, thus preventing differential potentials. Satellite builders go to considerable lengths to ensure this grounding, but some satellite surface devices, e.g. thermal radiators, are not amenable to this mitigation and can lead to charging problems.
- Higher energy electrons, particularly at MeV energies, can penetrate and deposit charge inside satellites. Where that charge is deposited inside dielectric materials such as electrical insulation and circuit boards, it can accumulate, eventually leading to discharges that can generate false signals in control systems, and even

cause damage to satellite systems. A vivid benchtop demonstration of such internal discharges is available online at https://youtu.be/eCz7BL74D4Y. Since these discharges occur in materials that must be non-conductive, they are more challenging to mitigate through design, except by provision of shielding to reduce the flux of electrons reaching vulnerable systems. Thus, satellite operators are generally ready to recognise and respond to anomalies caused by such discharges.

- High-energy (>10 MeV) ions can penetrate inside electronic systems on satellites, producing a trail of ionisation within devices such as processors, memory and power controllers. The trail of ionisation from a single ion is sufficient to disrupt device operation in what is termed a single-event effect (SEE). Examples of SEEs include the charge in the trail flipping a computer bit between 0 and 1 (single-event upset), the electric fields associated with the trail being misinterpreted as a real signal (single-event transient) or the trail triggering a burst of current that destroys the device (single-event burnout). As electronic devices become more sophisticated the range and diversity of SEEs has increased. See page 37 of Cannon et al. (2013) for a good summary of SEE types. SEEs can be mitigated by a variety of engineering methods, e.g. use of error correction codes to detect and fix upsets; use of three or more parallel decision circuits for control systems, such that a SEE in one circuit is outvoted by the other circuits. Nonetheless, SEEs are a fact of life for satellite operators, and one that they must always be ready to resolve.
- Another important example of the disruption that can be caused by high-energy particles is the generation of noise in sensors vital to satellite operations. A prime example is the star mappers that are very widely used to monitor satellite attitude by using stars as reference points. These devices quite obviously cannot be shielded, so high-energy particles can generate false signals in these sensors, and high fluxes of those particles can render star mappers unusable. A notable example occurred when ESA's Mars Express was in transit from Earth to Mars during the Halloween space weather event of 2003; the high fluxes of solar energetic particles blinded those mappers for some 15 h (NOAA 2004). Fortunately, this was not a serious risk during cruise phase, but would have been a major challenge if it had occurred close to important manoeuvres such as Mars orbit insertion.
- High-energy ions can also interact (via the Coulomb force) with the nuclei of materials inside satellite systems, causing the nuclei to be displaced from their proper positions in the material. This radiation damage will accumulate with time as satellites are exposed to these ions, leading to a gradual degradation in the properties of materials. Thus, the performance of systems such as solar arrays and electronic devices will gradually decline, eventually leading to those systems failing. Thus, this is an important constraint on satellite lifetime and an important consideration in satellite design. Satellite builders will seek to make systems sufficiently robust that the satellite will operate successfully over the lifetime, and within the cost, desired by their customer.
- Radiation damage is also an issue for biological materials on satellite—and hence for humans in space. The key concern here is long-term exposure to modest levels of radiation where damage to DNA increases the lifetime risk of cancer. This exposure is unavoidable (as it is certain working environments on Earth such as

nuclear industry, civil aviation and underground facilities) so mitigation is focused on keeping the exposure as low as reasonably possible (CDC 2015). Another concern is exposure to intense radiation events. These events bring the risk of high short-term exposure such that radiation damage to the human body causes sickness or even death. Clearly, it is essential to avoid such high levels of exposure, most obviously by sitting out the event in an area that is well sheltered from radiation (or if possible by evacuating to a safer area). Both these options exist for astronauts if an intense event occurs whilst in low Earth orbit: they can shelter in the best shielded part of their spacecraft, or evacuate by returning to Earth. Other options will be required on flights to Mars, as we discuss later in this chapter.

4.1.2 Other Space Weather Effects

We have so far focused on the space weather effects of energetic charged particles, such as those produced in solar radiation storms. But there are a number of other phenomena that produce space weather effects. Most importantly, there is the solar wind, the continuous flow of plasma away from the Sun, sometimes enhanced by the huge eruptions that we call coronal mass ejections (CMEs), as shown in Fig. 4.2. These outflows have little or no direct impact on satellites. The ram pressure of the solar wind is tiny, a 1000 times less than even the small radiation pressure delivered by sunlight. The solar wind only becomes important as a space weather phenomena when it interacts with plasma structures around other solar system objects—most obviously, the magnetosphere of the Earth (but also other large magnetospheres such as those around Mercury, Saturn and comets, and also mini-magnetospheres such as those found on the Moon and Mars). When the magnetic field in the solar wind has the opposite direction to that in a magnetosphere, solar wind energy and momentum can flow into the magnetosphere. For the Earth, this occurs when the solar wind magnetic field is southward, and the energy then injected can drive geomagnetic activity, leading to a wide range of space weather effects including the headline impact of power grid disruption, but also changes in Earth's upper atmosphere that can disrupt radio signals used for communications and navigation (e.g. GPS) as well as increased drag for satellites at altitudes below about 600 km (e.g. see Hapgood 2017a). These can be major effects because Earth's extensive magnetosphere provides a large collecting area (around 120000 km diameter) for solar wind and can focus that energy in time and space to produce intense effects in the upper atmosphere. This is, of course, not an issue for satellites in interplanetary space or even in orbit around Mars.

Another space weather effect that we should briefly consider is solar flares—brightenings on the surface of the Sun that produce intense fluxes of extreme ultraviolet (EUV) and X-ray radiation. This radiation has significant effects at Earth in that it causes a sharp rise of electron density at several levels in Earth's ionosphere. This can disrupt a number of radio technologies used on Earth. For example, GNSS systems will suffer a brief (tens of seconds) disruption as they adapt to the sudden change in the ionosphere, whilst high-frequency radio systems will suffer a more

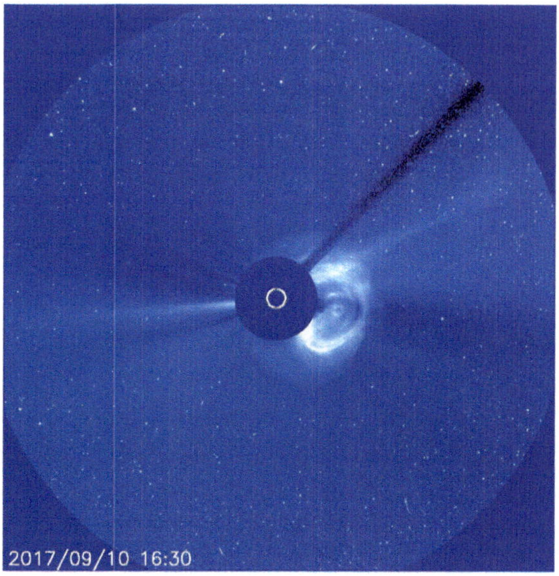

Fig. 4.2 Coronal mass ejection launched on 10 September 2017 appears in the right hand (west) side of this coronagraph image taken by the LASCO instrument on the ESA/NASA SOHO mission. The launch of this CME was associated with a radiation storm that penetrated to the surface of Mars, significantly raising radiation levels as measured by the RAD instrument on NASA's Curiosity rover (Hassler et al. 2018). Image courtesy NASA

extended (a few minutes to a few hours) loss of service due to the X-rays creating a layer of ionisation that absorbs high-frequency radio waves. But these technologies are not used in travel to Mars and their use on Mars is an issue for the distant future (Mars' ionosphere is likely to show similar effects). It is also important to note that intense fluxes of EUV and X-rays pose no direct risk to satellites as they are easily shielded by satellite surfaces. They may add slightly to the slow degradation of surfaces exposed to space, but that is just part of the normal "wear and tear" on satellites.

4.2 Space Radiation: Supernovae and Storms

As we have seen above, space weather poses a risk for Mars missions, mainly through the impact of the energetic charged particles that permeate interplanetary space. These arise from a number of sources, both inside and outside the solar system.

4.2.1 Galactic Cosmic Rays: Space Weather from Supernovae

We should first consider the sources outside the solar system, since these provide a slowly changing background with important long-term effects. These externally sourced particles are galactic cosmic rays (GCRs) and are thought to be energised

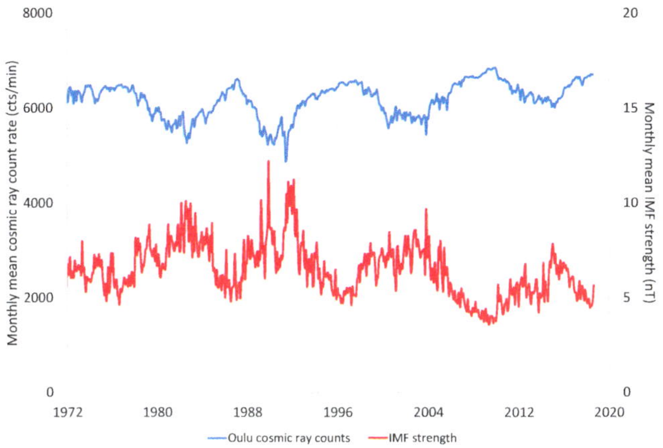

Fig. 4.3 Monthly mean values of the strength of the near-Earth interplanetary magnetic field (red), and the cosmic ray neutron count rate, as measured over solar cycles 21–24 (blue). The modulation of both quantities is clear with the IMF higher when the cosmic rate count rate is lower (both at solar maximum). IMF data from the NASA OMNIweb database (https://omniweb.gsfc.nasa.gov/) and neutron count rates from the cosmic ray station at Oulu in Finland (http://cosmicrays.oulu.fi/)

in shock waves around supernova explosions elsewhere in the galaxy. It is estimated that there are around two to three such explosions somewhere in our galaxy (i.e. the Milky Way) during each century (Tammann et al. 1994; Atri and Melott 2014) and that the resulting energetic particles are then trapped for thousands of years in the disc of our galaxy, in part by the magnetic fields that thread that disc (Parker 1969). The result is that interstellar space is filled by a fairly steady background of GCRs and that some of these particles enter the heliosphere, the region of space dominated by the solar wind, typically extending out to a distance around 70 astronomical units (Schrijver and Siscoe 2009). These GCRs will then be scattered by magnetic irregularities arising from turbulence in the solar wind, thereby reducing the flux of GCRs reaching the inner solar system. This reduction varies over the solar cycle with more scattering at solar maximum, when the interplanetary magnetic field is stronger, as shown in Fig. 4.3, and is more prone to irregularities. The scattering is particularly effective for GCRs with energies below a few GeV and so has a profound impact on the total GCR rate reaching the surface of the Earth (as also shown in Fig. 4.3), and will have an equally significant impact elsewhere in the inner solar system, not least missions to Mars.

As a result, GCRs provide a continuous, but low level, source of radiation for Mars missions, but nonetheless one with significant implication for mission design. There are several key issues.

a. it is very difficult to shield against GCRs because they include a significant flux of particles at very high energies (above 400 MeV). Thus, mission design should assume that some radiation will penetrate inside spacecraft during transit to, and in orbit around, Mars. This penetrating radiation from GCRs will lead to a low, but continuous, level of single-event events and radiation damage in spacecraft systems, so Mars missions must be designed to have high resilience to these effects.

b. the continuous (slowly changing but always present) GCR flux will lead to significant cumulative effects: most importantly, radiation dose in astronauts. Thus, the missions must be designed to monitor the radiation exposure of individual astronauts—as is done, and still being improved, on the International Space Station (Berger et al. 2016)—and to allow variations in astronaut daily routines so as to control that exposure.

c. there is a significant radiation environment in the atmosphere, and on the surface of, Mars. This is similar to the radiation environment at high altitudes in Earth's atmosphere, in particular, high fluxes of secondary particles (such as neutrons and muons) produced by collisions between GCRs and atmospheric species. At Earth this radiation environment is a significant space weather issue for aviation and is the subject of extensive studies (Cannon et al. 2013; Meier and Matthiä 2014; Dyer et al. 2018). At Mars it is also a focus for both theoretical (Dartnell et al. 2007) and observational studies (Hassler et al. 2018)—and for practical measures in terms of the resilience of systems on-board Mars rovers (Laird et al. 2007; Kolawa et al. 2013). It is clear that any long-term human presence on Mars' surface will require that accommodation is buried, either by digging down into the regolith or by building a mound of regolith material over the accommodation. Thus, an essential piece of kit for human exploration will be a mechanical digger, a Mars-qualified version of the machines seen on building sites across Earth.

4.2.2 *Solar Radiation Storms*

The Sun is also a source of very high-energy particles, but a highly episodic one as bursts of activity in the Sun's atmosphere send bursts of energetic particles out into the solar system. Figure 4.4 shows an example of these solar radiation storms as observed at Earth. As you can see the bursts can give a thousand-fold or more increase in the fluxes of energetic particles at MeV energies (and sometimes at GeV energies). Thus, they can greatly increase the risks from several of the effects discussed above including single-event effects, radiation damage to spacecraft systems and the radiation exposure of astronauts.

These particles in these storms are thought to be energised either at the shock waves in front of fast coronal mass ejections (Reames 1999) or when the magnetic fields in the Sun's atmosphere are reconfigured, a process that we call magnetic reconnection (Reames 1999; Drake et al. 2009). The latter is a fundamental plasma process in which complex magnetic fields change to a simpler topology, and as a side effect, convert magnetic energy into kinetic energy. The solar energetic particles (SEPs) energised at CME shocks will almost inevitably escape into interplanetary space since those shocks are magnetically connected to the solar wind, but the escape of those energised by reconnection will depend on the strength and topology of the

Fig. 4.4 Solar energetic particle fluxes during a major radiation storm at Earth in October 1989, perhaps the most sustained SEP event since the advent of spaceflight. The storm was generated and sustained by at least three major bursts of activity on the surface of the Sun (indicated by the red arrows). The storm exceeds NOAA's S3 level (https://www.swpc.noaa.gov/noaa-scales-explanation) for around 100 h. Similar conditions during a flight to Mars could require astronauts to shelter in a well-shielded area and could disrupt operation of many on-board systems. The feature indicated by the black arrow is discussed in Fig. 4.5. Data are >10 MeV proton fluxes derived from measurements by the GME instrument on NASA's IMP-8 satellite, which was then in orbit around Earth at a distance around 35 Re. At this distance, the satellite is almost fully exposed to SEP fluxes, even when in the magnetosphere (period indicated by green bar). *Data Source* NASA Space Physics Data Facility

magnetic fields overlaying the reconnection site. They will escape only if those overlying fields are weak or have a topology that links to the solar wind.

The existence of these different SEP energisation processes leads to considerable diversity in the nature of solar radiation storms.

a. radiation storms produced by reconnection are necessarily short-lived (hours) because the reconnection process is itself short-lived. Thus, energisation by reconnection is thought to be the origin of radiation storms lasting a few hours up to a day, those termed abrupt by Reames (1999). (The events are also associated with solar flares as the reconnection process also generates the intense electron fluxes that collide with dense gas at the surface of the Sun, thereby producing solar flares.)

b. radiation storms produced by fast CME shocks can last for days as the shocks propagate out into the solar wind, and were termed gradual by Reames (1999).

c. shocks and other magnetic structures in the solar wind can trap particles with energies below about 50 MeV, leading to sharp increase in the flux of particles at these energies when the structure passes over a spacecraft (see Fig. 4.5). These "energetic storm particles" can have significant effects on exposed spacecraft systems such as degradation of solar arrays, noise in external sensors and, of course, radiation doses for any astronauts working outside a spacecraft.

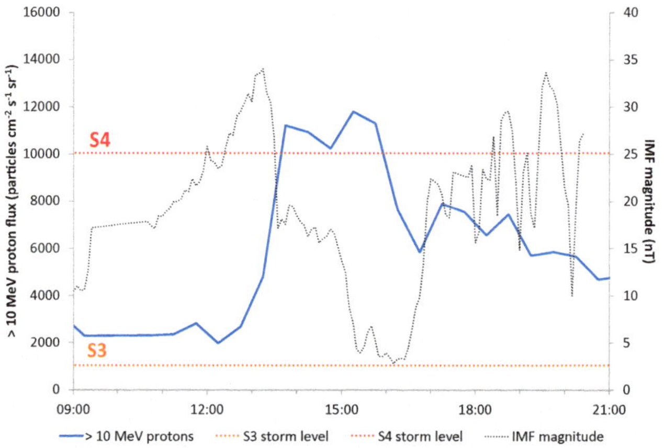

Fig. 4.5 An energetic storm particle (ESP) event observed at Earth on 20 October 1989. This is an expanded view of the event marked by the black arrow in Fig. 4.4. During this event, the SEP fluxes are enhanced by almost an order of magnitude for several hours, lifting the radiation storm above NOAA's S4 level. These high SEP fluxes are trapped in a region of low magnetic field strength as shown by the dotted black curve (and also low plasma density, not shown in the figure). This is typical of ESP events, very high fluxes of 1–50 MeV particles trapped in solar wind structures. They could be a serious hazard during a flight to Mars as the high fluxes can degrade solar arrays and also deliver significant radiation dose to any astronauts working outside well-shielded area. Data are >10 MeV proton fluxes from IMP-8 as in Fig. 4.4, and interplanetary magnetic field data from NASA's OMNIweb database. *Data Source* NASA Space Physics Data Facility

d. the potential for false alarms of radiation storms when strong overlying magnetic fields prevent the escape of SEPs. There are instances when a significant activity can occur in the Sun's atmosphere, but no SEPs (or CMEs) escape into interplanetary space. This was well demonstrated when a number of large solar flares occurred in October 2014, but there was no significant ejection of plasma or particle radiation from the Sun.

One important feature of solar radiation storms is that they affect only part of interplanetary space, as illustrated in Fig. 4.6. The SEPs will flow into the region that is magnetically connected to the source region in the Sun's atmosphere or to the CME shock as it propagates out into interplanetary space. Thus, the affected region follows the topology of the interplanetary magnetic field (IMF), namely a spiral pattern known as the Parker spiral (Parker 1958). (This pattern arises naturally as the magnetic flux tubes that form the IMF are pulled out of the Sun by the radial flow of the solar wind (Hansteen 2009) but the footprint of each flux tube remains attached to the rotating Sun.) The width of the affected region will gradually increase with time during a storm as some particles are scattered perpendicular to the IMF. (Also, the spiral pattern of the IMF and of the affected region may be distorted by the propagation of CMEs through the region.) But the critical issue is that the spatial extent of a radiation storm is constrained by the topology of the IMF. Thus, any particular radiation storm may affect Earth and Mars quite differently, depending on

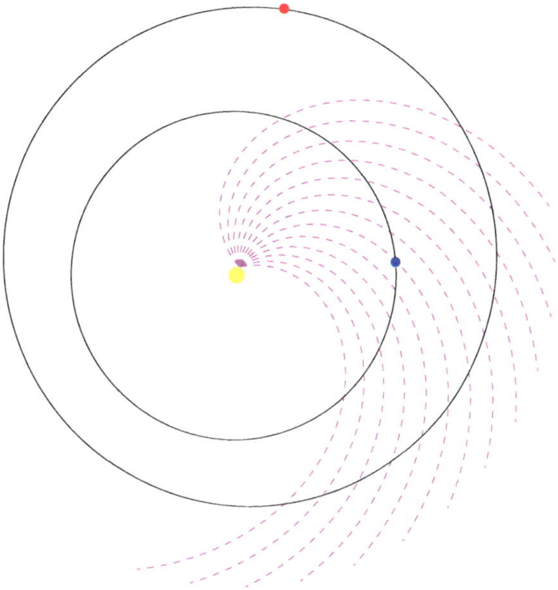

Fig. 4.6 Schematic showing how solar energetic particles follow the Parker spiral of the interplanetary magnetic field (magenta dashed lines) to fill a large fraction of interplanetary space. In this schematic Earth (blue) is enveloped by the SEP stream but Mars (red) is not impacted. The black ellipses show the orbits of Earth and Mars using standard Keplerian elements (BAA 2017). The Parker spiral is shown for a solar wind speed of 400 km s^{-1}

the relative ecliptic longitudes of the two planets. They are most likely to experience similar conditions a few months after the Mars opposition, so that Mars is lagging Earth by around 30°, as shown in Fig. 4.7. However, when Mars is around conjunction, the two planets are likely to experience very different conditions. As we will see below, this has a profound impact on the requirements for forecasting space weather conditions that could affect astronauts on a Mars mission.

Solar radiation storms often include a significant flux of very high-energy electrons that have the potential to cause charging on exposed bodies. This has been demonstrated on the Moon, where measurements by NASA's Lunar Prospector mission showed that the lunar surface was charged up potentials of several thousand volts during radiation storms (Halekas et al. 2007). Thus, it is possible that radiation storms could cause significant surface charging on spacecraft bound travelling to/from Mars. Whilst this has not, so far, been a problem for robotic spacecraft, it is a risk that needs a study for human missions, especially if those missions envisage that astronauts may perform maintenance tasks outside the spacecraft.

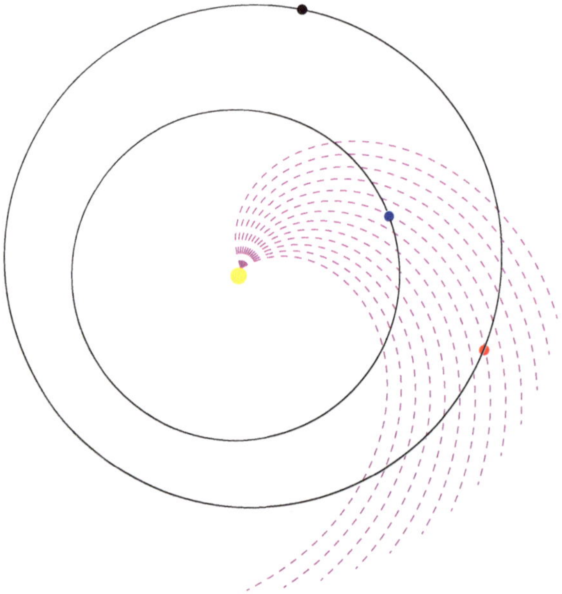

Fig. 4.7 Schematic showing a planetary alignment in which Earth (blue) and Mars (red) will both lie in the centre of a SEP stream (magenta dashed lines) flowing away from the Sun. Since the stream follows the Parker spiral of the interplanetary magnetic field, Mars will then lie around 30° behind the Earth in its orbit around the Sun (the planets move anticlockwise in this view). The SEP source region on the Sun will lie around 60° ahead of the Earth and hence will be best observed by a spacecraft at the longitude indicated by the black circle. As in Fig. 4.6, the black ellipses show the orbits of the two planets using standard Keplerian elements (BAA 2017). The Parker spiral is again shown for a solar wind speed of 400 km s^{-1}

4.3 So What Do Astronauts Need?

4.3.1 Introduction: Do We Need a Mixed Architecture?

As we have shown above, the major space weather hazard for astronauts travelling to and from Mars comes from energetic particles: (a) in terms of the radiation dose that they will accumulate during travel; (b) in terms of how particles may disrupt the operation of spacecraft systems (through single-event effects, noise in sensors, and charging); and (c) through the ageing of spacecraft systems through radiation damage. Some of these effects will also be significant when working on the surface of Mars. Figure 4.8 shows a simple practical example: an image of Mars' night sky taken by the NASA Spirit rover over a 5-minute period on 11 March 2004; stars appear as parallel near-vertical streaks due to the rotation of Mars in those 5 minutes, whilst radiation hits in the image sensor appear as points or short streaks running at an angle to star tracks. Solar activity was low at the time of this image, so we can conclude that, in this case, the radiation arose from galactic cosmic rays.

Fig. 4.8 Five-minute exposure of the night sky on Mars as NASA's Spirit rover on 11 March 2004. Stars appear as parallel long near-vertical streaks (due to the rotation of the planet) against a background of points or short streaks arising from radiation hits. Image courtesy of NASA, *Source* http://photojournal.jpl.nasa.gov/catalog/PIA05551

As noted above, the planning of human travel to Mars needs to consider both: (a) background fluxes of very energetic particles in the form of galactic cosmic rays and (b) episodic high fluxes of energetic particles that occur during solar radiation storms. This mix of sources is fundamental to mission architecture. The background radiation is present throughout any journey to Mars, including time in orbit around that planet, and time spent on its surface. Thus, any Mars mission must be continuously capable of dealing with problems caused by GCRs, e.g. occasional single-event effects, gradual degradation of systems due to radiation damage and gradual accumulation of radiation dose by astronauts. The whole mission design must incorporate engineering measures to mitigate background SEE rates in all spacecraft systems, as well as shielding to limit cumulative radiation effects. However, it is not essential that those measures be designed to a level that can also mitigate occasional solar radiation storms, where we expect much higher SEE rates and a radiation risk to astronauts. There is potential to mitigate these by special measures: by astronauts taking shelter in a more heavily shielded area that is equipped to sustain them for the duration of the storm (a few days), and that includes remote controls that enable astronauts to monitor and reset critical systems. In this case the mission could ride the storm—handling high SEE rates through human action (much as is done today by satellite operators on the ground) and recognising that storms make only a modest contribution to cumulative radiation damage in satellite systems (in general, the GCR background makes a bigger contribution Schwadron et al. 2014).

The key requirement here is to protect the astronauts, so the mission architecture must reflect a decision on which of two approaches is the more reasonable way to minimise radiation effects on astronauts: (a) should the architecture include a large radiation shield that can protect most of the spacecraft including all living space and critical systems? Or (b) should it include a well-equipped shelter to which astronauts can retreat for the few days duration of a radiation storm.

The first approach will require extensive shielding that protects all of the on-board living space and all of the critical systems. One option for this shielding is a large volume of material that will scatter incoming particles through interactions with the nuclei of atoms in that material. Traditional spacecraft radiation shielding is composed of high mass density materials such as metals due to limited space on spacecraft. But a Mars mission will very likely require a large spacecraft assembled in low Earth orbit. This offers the opportunity to build physically large radiation shields using materials containing low mass nuclei (and typically with low mass density). This offers two important advantages: (a) collisions with low mass nuclei are more effective at reducing the energy of incoming particles and (b) those collisions produce less secondary radiation. A number of low mass materials have been considered as candidate materials for physically large radiation shields, including polythene (Barthel and Sarigul-Klijn 2017; Barthel and Sarigul-Klijn 2018), water (Adamo and Logan 2016; Vanatta et al. 2018) and liquid hydrogen (Vanatta et al. 2018). The latter options also open up the possibility for dual use of spacecraft resources whereby shielding materials can also serve other purposes such as astronaut water supplies or fuel for propulsion. Further studies are needed to deepen our understanding of the pros and cons of different shielding materials and in the case of dual use, of the potential trade-offs with other mission requirements. Another shielding option is to surround the spacecraft with a mini-magnetosphere similar to those found around the magnetic anomalies on the Moon (Lin et al. 1998) and that protect parts of the lunar surface from space weathering by deflecting incoming particles (Bamford et al. 2012). If a similar mini-magnetosphere were created around a spacecraft (e.g. through use of powerful superconducting magnets), it can be filled with a cloud of plasma that can deflect incoming particles by Coulomb interactions with ions in that cloud (Bamford et al 2014). Thus, one could create a particle scattering field around the spacecraft, and hence reduce the flux of radiation impinging on the astronauts and on vulnerable spacecraft systems. This plasma shield would also have the advantage of not producing secondary particle radiation—since it uses the Coulomb force, and not nuclear forces, to deflect particles.

The second approach will require that only part of the spacecraft is heavily shielded, providing an area that is just large enough for the astronauts to sit out the storm and that is equipped to sustain them for a few days, e.g. with life support, communications, and of course, radiation monitoring. It would likely be embedded deep inside the spacecraft, so other spacecraft systems contribute to the shielding, in addition to shielding immediately around the shelter (e.g. see Alenia-Aerospazio 2004). In this situation, we would need to use mass shielding: metal as in contemporary spacecraft or by use of light materials as discussed above. For example, the shelter could be surrounded by a large water tank inside the spacecraft.

4.3.2 Exploiting Nature's Shielding

In addition to the shielding provided by the spacecraft, Mars missions can also reduce their exposure to radiation from galactic cosmic rays by picking the right time to travel to Mars. As previously discussed the flux of cosmic rays varies markedly with solar activity, particularly the solar cycle, as shown in Fig. 4.3. Thus, the best time to send a human mission to Mars is during the years around solar maximum (Schwadron et al. 2014), when the turbulent magnetic fields in the solar wind are most effective as a natural radiation shield that scatters GCRs. Travel to Mars around solar maximum will reduce both the occurrence of problems, and the accumulation of damage caused by background radiation. There will, of course, still be a need to mitigate the radiation storms that are more common around solar maximum, e.g. through use of a shelter as discussed above.

Unfortunately, the 11-year length of the solar cycle is a significant constraint on how we can exploit the natural shielding linked to high solar activity. That better shielding is limited to a period of 5–7 years around solar maximum, leaving 4–6 years around solar minimum when cosmic ray fluxes will be naturally higher. This is not a big issue for a short campaign of human missions to Mars on the lines of the Apollo missions to the Moon, and even less of an issue for a one-off mission. These missions can be planned to take place during solar maximum. But it will be a big issue if and when there is a sustained long-term programme of Mars exploration. The programme will have to make major decisions on how to schedule flights to Mars—perhaps with human travel being focused into solar maximum periods, complemented by robotic supply missions that could take place at any phase of the solar cycle. Continuous human presence on Mars might then require that the astronauts remain on Mars for several years during the solar minimum period, returning only when rising solar activity reduces cosmic ray fluxes. This approach would be similar to over-wintering of staff at Antarctic research bases as practised over the past 60 years, but requiring a much longer stay in the case of Mars (though with better options for resupply than is usually done in Antarctica, even today).

The use of this natural shielding may also be constrained by the downward trend of solar activity that has become apparent in recent years, e.g. as shown by the weak maximum of solar cycle 24 (McComas et al. 2013). This downward trend is thought to be a continuation of longer term variations in which solar activity rises and falls over periods of several hundred years. Studies of cosmogenic isotopes (which are a proxy for cosmic ray fluxes) show that over twenty such cycles have occurred over the past 9000 years (Usoskin 2017). These cycles include extended periods of very low solar activity (such as the Maunder Minimum of the seventeenth century) and of very high solar activity (as in the mid- and late twentieth century). The present downward trend suggests that solar activity may be low during much of the twenty-first century or perhaps even into twenty-second century (Lockwood et al. 2011; Steinhilber and Beer 2013), thus reducing the duration of the periods that are favourable for human travel to Mars (Schwadron et al. 2014, 2018).

These trends suggest, whilst we should exploit whatever natural shielding is provided by magnetic irregularities in the solar wind, we should design human missions to Mars to have a high resilience to the background radiation from cosmic rays. Designs should assume radiation conditions significantly higher than experienced during the past 60 years of space exploration, at least twice as large to reflect estimates of the cosmic ray fluxes reaching Earth during the low solar activity of the Maunder Minimum (McCracken et al. 2004; Muscheler et al. 2007), and perhaps even higher to allow for even lower solar activity, as suggested by the historical record (Usoskin 2017).

4.3.3 Coping with Storms

As we have seen there are a wide range of ways to build space weather resilience into human missions to Mars. Good engineering can harden spacecraft systems and reduce the radiation doses accumulated by astronauts. It is also worth noting that the presence of astronauts will add to the resilience of spacecraft systems, since they will be much better placed than ground controllers to diagnose problems and reset systems. Nonetheless, all these measures will be challenged when adverse space weather leads to a major solar radiation storm. Thus, it is vital that astronauts and ground controllers are well aware of when a radiation storm is imminent and when it is ongoing. That awareness will enable prompt action to deal with the storm, e.g. astronauts moving to the best-protected areas of their spacecraft (and to equivalent in any habitat on the surface of Mars). Ideally, this would be an area specifically designed as a radiation shelter, but could just be the area that is best protected due to other design considerations.

Thus, we must consider how astronauts and ground controllers can gain awareness of adverse space weather conditions. What elements of space weather monitoring and forecasting are required? First of all, we must note that the speed of light prevents real-time communications between the astronauts and ground controllers. The one-way-light-time between Earth and a Mars mission can be up to 20 minutes. This is much too long for ground control to help astronauts detect the onset or enhancement of a radiation storm (note the four abrupt increases in radiation flux marked by arrows in Fig. 4.4). So it is essential that spacecraft carrying humans have an autonomous capability to detect the high-energy particles that form radiation storms. In contrast, ground control can play a major role in advising astronauts of when conditions on the Sun suggest that a radiation storm is imminent—and that storm likely to impact the astronauts and their spacecraft. This forecasting function will require capabilities that cannot be placed on the astronauts' spacecraft, in particular, remote sensing of conditions on the Sun (best done on separate dedicated spacecraft) and assessment of those conditions by expert space weather forecasters (best done on Earth).

It is worth discussing what actions the astronauts could take in response to a forecast of an imminent radiation storm. Since radiation is a cumulative risk, not an immediate risk, they can adopt a phased approach in which they make themselves ready to shelter from the storm, but move into the shelter only when the storm onset is detected via on-board sensors. This will allow them to avoid unnecessary sheltering for long periods. Instead, the forecast can trigger action to ensure that the shelter is ready, i.e. well stocked with food and water, and any other items needed whilst sheltering, including anything needed to maintain morale (basically items similar those needed whilst sheltering from any threat here on Earth). The astronauts can also take action to constrain their working activities—to focus on activities that can be stopped quickly (a few minutes at most), so that astronauts can shelter promptly when the storm onset is detected. The action of moving into the shelter may be enhanced if the on-board sensors can respond to the fastest-arriving particles in a radiation storm. These are typically MeV electrons, which travel more quickly than high-energy ions because of their low mass (Posner 2007).

The response to storm onset will be a little more complicated for humans working on the surface of Mars, since their local radiation monitors will detect only the radiation that penetrates Mars atmosphere. This points to the value, indeed importance, of maintaining a radiation monitoring capability in orbit around Mars to enable prompt detection of events, and of good real-time communications between that capability and any humans working on the surface (the distances involved are short so there are no speed of light limitations).

4.3.4 Space Weather Forecasting: What Do Mars Missions Need?

It is important to remember here that the particles in a radiation storm follow the spiral trajectory of the interplanetary magnetic field, e.g. as shown in Fig. 4.7. So, the solar source of a high-risk storm is not easily viewed from the astronaut's spacecraft. It is this displacement in longitude that drives the requirement for space weather monitoring on a separate spacecraft. A Mars mission will require the capability to monitor space weather conditions in the solar regions that are magnetically connected to the astronauts' spacecraft—and hence, a spacecraft displaced in longitude as shown by the black circle in Fig. 4.7. The monitoring spacecraft will need, at the very least, to carry instruments that can observe the development of magnetic complexity in the Sun's atmosphere. That complexity, often in the form of structures around sunspot groups, is an indication of the buildup of magnetic energy in the Sun's atmosphere, an essential precursor to a radiation storm (and to other space weather effects). Thus, this is essentially a requirement to place a standard space weather remote sensing capability over the solar regions that can launch a radiation storm towards the astronauts.

For astronauts working on Mars, that solar region will be around 90° ahead of Mars in its orbit (assuming that the storm particles follow the Parker spiral of the interplanetary magnetic field, the difference in longitude may be estimated as $\Delta\varphi = R\Omega/v_s$, where R is the distance of Mars from the Sun, Ω is the angular velocity of rotation of the Sun, 2.7×10^{-6} rad s^{-1} and v_s is the solar wind speed, typically 400 km s^{-1}). However, it would be difficult to maintain a stable orbit at 90° ahead. As a compromise location, one may consider placing a monitoring spacecraft at the Lagrange L4 point, some 60° ahead of Mars in its orbit. But that would require some consideration of collision risks with other objects around that L4 point: at least one asteroid (1999 UJ7) is known to be in a stable orbit around that point (Scholl et al. 2005).

For astronauts travelling from Earth to Mars, the situation is more complex because the solar source region will vary from 60° ahead of the spacecraft at launch from Earth and gradually move to 90° ahead as the spacecraft travels to Mars. The situation will be reversed for the return journey to Earth. It would be challenging to design orbits so that a spacecraft can monitor these solar regions in phase with the motion of the astronauts' spacecraft. At the very least, it would almost certainly require separate monitoring missions for the outward and return journeys. Also, there would be a significant risk that launch delays could degrade the relative phasing of the spacecraft carrying the astronauts and the monitoring instruments.

In summary, it would be cumbersome to provide a space weather monitoring spacecraft on a dedicated basis, tied to the motion of astronaut spacecraft and of Mars. It would be much more effective to build a comprehensive system for global monitoring of space weather conditions on the Sun, in particular, a system to provide continuous full-Sun monitoring of solar magnetic field structures. Monitoring across all 360° of solar longitude will ensure that we can detect developing space weather threats for any phase of a Mars mission, even when the mission is behind the Sun as seen from Earth. This approach also has a major advantage, and in that, it will be synergistic with requirements arising from other space weather forecasting services, in particular, those that serve the needs of Earth-bound users. For example, full-Sun monitoring has been identified as an important future objective for wider space weather research and operations (Schrijver et al. 2015) as it is an essential element in building reliable global models of the evolving solar magnetic fields that are the main energy source for space weather. Partial monitoring of solar magnetic fields can lead to a range of problems, e.g. unexpected changes due to long-range linkages with hidden magnetic field features.

This full-Sun monitoring will require a ring of satellites around the Sun and has already been subject of mission concept studies (e.g. Ritter et al 2015). Many of the major elements needed for these satellites have already been demonstrated or are under study as part of wider work on space weather. For example, the solar and heliospheric imagers on NASA's STEREO mission (Howard et al. 2008) have already demonstrated the space weather value of having a viewpoint away from the Sun-Earth line (Harrison et al. 2017). Unfortunately, STEREO did not carry a magnetograph so the exploitation of STEREO data has focused on other space weather issues such as multi-viewpoint observations of CMEs (e.g. Harrison et al. 2017). But STEREO has stimulated many ideas for follow-up missions that enable us to review regions of the Sun hidden from the Earth. In particular, there has been much interest in missions to the Lagrange L5 point (e.g. Vourlidas 2015; Hapgood 2017b) as this viewpoint will give us better views of regions of the Sun approaching the Earth-Sun line (and hence prime sources of space weather that could affect Earth). But the requirements of a single mission to L5 are very similar to those needed for each individual spacecraft in our proposed observing ring around the Sun: we need the same instruments to observe the Sun and to monitor local conditions in the solar wind flowing past the spacecraft, we also face the same issues in terms of returning sufficient high-quality data to Earth over distances of several hundred million kilometres.

Thus, it is timely to note that the European Space Agency (ESA) is now leading the design of space weather mission to L5 under its Lagrange project, with the aim of flying the mission by the mid-2020s (Gibney 2017; Luntuma et al. 2018). The primary goal of Lagrange is very practical—namely to improve forecasts of space weather at Earth, thereby helping the ESA member states, and their global partners, to manage the risk that space weather poses to critical infrastructures. But it is perhaps time to recognise that Lagrange can also act as a pathfinder for the space weather monitoring needed to support human exploration of Mars. Our proposed observing ring is essentially a set of 6–10 Lagrange spacecrafts spread around the Sun. The instruments are the same, the thermal and radiation environments will be similar (assuming the ring is 1 AU from the Sun) and the challenge of data downlink is broadly similar. Indeed, ESA is studying the use of optical communications for interplanetary missions (Sodnik et al. 2017); in the long-term this has the potential to greatly improve the return of high-quality data (e.g. high-resolution images) from future space weather missions including our proposed observing ring.

4.4 Summary and Conclusions

Space weather poses a risk to the human exploration of Mars mainly through solar radiation storms. These arise when charged particles are accelerated to high energies through the release of magnetic energy stored in structures within the Sun's atmosphere. These high-energy particles can penetrate into spacecraft where high-energy ions can disrupt electronic systems through single-event effects and high-energy electrons can deposit electrical charge inside dielectric materials, possibly leading to disruptive electrical discharges. In addition exposure to high-energy ions can gradually damage the structure of materials in satellite systems and in the bodies of astronauts, leading to a deterioration in the performance of satellite systems and increased cancer risk for astronauts.

Thus, the critical space weather issue is to mitigate the risk that radiation poses to Mars missions, both the risk to astronauts and the risk to spacecraft systems. This mitigation needs to be integrated with the mitigation of the radiation risk from galactic cosmic rays. GCRs provide a slowly changing background flux of radiation that dominates the cumulative radiation exposure of astronauts and spacecraft systems, whilst radiation storms provides bursts of much higher fluxes but each burst lasting only a few days. A human mission to Mars must be designed to mitigate both features. It must have shielding and other radiation hardening that can provide continuous mitigation of GCRs: keeping the astronaut radiation exposure as low as reasonably practicable and using good engineering design to minimise disruption of spacecraft systems. Ideally, it would apply similar measures to deal with radiation storms, but in practice, it may be better to provide a well-equipped shelter to which astronauts can retreat during radiation storms.

This implies that the astronauts, and their support teams back on Earth, need good situational awareness of the space weather conditions facing the mission. Most importantly the spacecraft will need a good (and resilient) set of sensors that can detect the onset of a radiation storm and monitor its progress. This is essential to enable the astronauts to make informed decisions about when to enter and leave the shelter. But this local monitoring of space weather conditions should be complemented by forecasts provided by experts back at Earth. Such forecasting will require monitoring of magnetic activity in all longitude zones of the Sun's atmosphere, not just the 180° of longitude that can be seen from Earth (of which perhaps the central 120° is well observed). This monitoring will require a ring of space weather monitoring spacecraft surrounding the Sun. The requirements for these spacecraft are very similar to those for missions now being proposed to monitor Earth-directed space weather from viewpoints off the Earth-Sun line, notably the ESA Lagrange project. Thus, it is timely to recognise the potential for synergy between the space weather requirements for Mars mission and those for forecasting space weather at Earth. The concepts and missions now being developed to advance those forecasts can also act as pathfinders for a system to support Mars missions. Indeed, looking further ahead, an integrated system for monitoring space weather conditions across the whole of the Sun's atmosphere could serve both needs.

References

Adamo, D. R., & Logan, J. S. (2016). Aquarius, a reusable water-based interplanetary human spaceflight transport. *Acta Astronautica, 128,* 160–179. https://doi.org/10.1016/j.actaastro.2016.07.026.

Atri, D., & Melott, A. L. (2014). Cosmic rays and terrestrial life: a brief review. *Astroparticle Physics, 53,* 186–190. https://doi.org/10.1016/j.astropartphys.2013.03.001.

Alenia-Aerospazio. (2004). Alenia-Aerospazio. Space Division, REMSIM-Final Report. Report to ESA ALS-RMS-VRP-001, 15 Sept. 2004.

BAA. (2017). Handbook of the British Astronomical Association: 2018. ISSN 0068-130-X.

Bamford, R. A., Kellett, B., Bradford, W. J., Norberg, C., Thornton, A., Gibson, K. J., et al. (2012). Minimagnetospheres above the lunar surface and the formation of lunar swirls. *Physical Review Letters, 109*(8), 081101. https://doi.org/10.1103/physrevlett.109.081101

Bamford, R. A., Kellett, B., Bradford, J., Todd, T. N., Benton, M. G., Stafford-Allen, R., et al. (2014). An exploration of the effectiveness of artificial mini-magnetospheres as a potential solar storm shelter for long term human space missions. *Acta Astronautica, 105,* 385–394. https://doi.org/10.1016/j.actaastro.2014.10.012.

Barthel, J., & Sarigul-Klijn, N. (2017). Adaptive, readily morphing, optimized radiation shielding for transit habitats: Flyby Mars mission. *Journal of Spacecraft and Rockets, 54,* 1196–1201. https://doi.org/10.2514/1.A33935.

Barthel, J., & Sarigul-Klijn, N. (2018). Radiation production and absorption in human spacecraft shielding systems under high charge and energy Galactic Cosmic Rays: Material medium, shielding depth, and byproduct aspects. *Acta Astronautica, 144,* 254–262. https://doi.org/10.1016/j.actaastro.2017.12.040.

Berger, T., Przybyla, B., Matthiä, D., Reitz, G., Burmeister, S., Labrenz, J., et al. (2016). DOSIS & DOSIS 3D: long-term dose monitoring onboard the Columbus Laboratory of the International Space Station (ISS). *Journal of Space Weather and Space Climate, 6,* A39. https://doi.org/10.1051/swsc/2016034

Cannon, P., et al. (2013). Extreme space weather: Impacts on engineered systems and infrastructure. UK Royal Academy of Engineering, London, ISBN 1-903496-95-0.

CDC (Centers for Disease Control and Prevention). (2015). ALARA-As low as reasonably achievable. Retrieved July 29, 2018, from https://www.cdc.gov/nceh/radiation/alara.html.

Dartnell, L. R., Desorgher, L., Ward, J. M., & Coates, A. J. (2007). Modelling the surface and subsurface Martian radiation environment: Implications for astrobiology. *Geophysical Reseach Letters, 34,* L02207. https://doi.org/10.1029/2006GL027494.

Drake, J. F., Cassak, P. A., Shay, M. A., Swisdak, M., & Quataert, E. (2009). A magnetic reconnection mechanism for ion acceleration and abundance enhancements in impulsive flares. *The Astrophysical Journal Letters, 700,* L16. https://doi.org/10.1088/0004-637X/700/1/L16.

Dyer, C., Hands, A., Ryden, K., & Lei, F. (2018). Extreme atmospheric radiation environments & single event effects. *IEEE Transactions on Nuclear Science, 65*(1), 432–438. https://doi.org/10.1109/TNS.2017.2761258.

Gibney, E. (2017). Space-weather forecast to improve with European satellite. *Nature, 541,* 271. https://doi.org/10.1038/541271a.

Halekas, J. S., Delory, G. T., Brain, D. A., Lin, R. P., Fillingim, M. O., Lee, C. O., et al. (2007). Extreme lunar surface charging during solar energetic particle events. *Geophysical Reseach Letters, 34,* L02111. https://doi.org/10.1029/2006GL028517.

Hapgood, M. (2017a). *Space weather*. IOP Publishing, Bristol. ISBN 978-0-7503-1372-8. https://doi.org/10.1088/978-0-7503-1372-8

Hapgood, M. (2017b). L1L5Together: Report of workshop on future missions to monitor space weather on the sun and in the solar wind using both the L1 and L5 lagrange points as valuable viewpoints. *Space Weather, 15,* 654–657. https://doi.org/10.1002/2017SW001652.

Hassler, D. M., Zeitlin, C., Ehresmann, B., Wimmer-Schweingruber, R. F., Guo, J., Matthiä, D., et al. (2018). Space weather on the surface of Mars: Impact of the September 2017 events. *Space Weather, 16*. https://doi.org/10.1029/2018sw001959

Hansteen, V. H. (2009). Stellar winds and magnetic fields. In C. J. Schrijver, & G. L. Siscoe (Eds.), *Heliophysics: Plasma physics of the local cosmos*. Cambridge University Press. ISBN-13:978-1107403222

Harrison, R. A., Davies, J. A., Biesecker, D., & Gibbs, M. (2017). The application of heliospheric imaging to space weather operations: Lessons learned from published studies. *Space Weather, 15,* 985–1003. https://doi.org/10.1002/2017SW001633.

Howard, R. A., Moses, J. D., Vourlidas, A., Newmark, J. S., Socker, D. G., Plunkett, S. P., et al. (2008). Sun Earth connection coronal and heliospheric investigation (SECCHI). *Space Science Reviews, 136*(1–4), 67. https://doi.org/10.1007/s11214-008-9341-4.

Kolawa, E., Chen, Y., Mojarradi, M. M., Weber, C. T., & Hunter, D. J. (2013). A motor drive electronics assembly for Mars Curiosity Rover: An example of assembly qualification for extreme environments. In *2013 IEEE Reliability Physics Symposium*, Anaheim, CA, 2013, pp. 2E.2.1–2E.2.9. https://doi.org/10.1109/irps.2013.6531963.

Laird, J. S., Scheick, L., Vizkelethy, G., Mojarradi, M. M., Miyahira, T., Chen, Y., et al. (2007). Single event transient analysis of an SOI operational amplifier for use in low-temperature Martian exploration. *Nuclear Instruments and Methods in Physics Research Section B: Beam Interactions with Materials and Atoms, 260*(1), 281–287.

Lin, R. P., Mitchell, D. L., Curtis, D. W., Anderson, K. A., Carlson, C. W., McFadden, J., et al. (1998). Lunar surface magnetic fields and their interaction with the solar wind: Results from Lunar Prospector. *Science, 281*(5382), 1480–1484. https://doi.org/10.1126/science.281.5382.1480.

Livadiotis, G., & McComas, D. J. (2013). Understanding kappa distributions: A toolbox for space science and astrophysics. *Space Science Reviews, 175,* 183–214. https://doi.org/10.1007/s11214-013-9982-9.

Lockwood, M., Owens, M. J., Barnard, L., Davis, C. J., & Steinhilber, F. (2011). The persistence of solar activity indicators and the descent of the Sun into Maunder Minimum conditions. *Geophysical Reseach Letters, 38,* L22105. https://doi.org/10.1029/2011GL049811.

Luhmann, J. G., Zeitlin, C., Turner, R., Brain, D. A., Delory, G., Lyon, J. G., et al. (2007). Solar energetic particles in near-Mars space. *Journal Geophysical Research, 112,* E10001. https://doi.org/10.1029/2006JE002886.

Luntuma, J. P., Kraft, S., & Glover, A. (2018). ESA SSA Lagrange mission to L5 point. Retrieved August 22, 2018, from https://cpaess.ucar.edu/sites/default/files/documents/sww-2018-presentations/Luntama_Juha-Pekka_06.pdf.

McComas, D. J., Angold, N., Elliott, H. A., Livadiotis, G., Schwadron, N. A., Skoug, R. M., et al. (2013). Weakest solar wind of the space age and the current "mini" solar maximum. *The Astrophysical Journal, 779*(1), 2. https://doi.org/10.1088/0004-637X/779/1/2.

McCracken, K. G., McDonald, F. B., Beer, J., Raisbeck, G., & Yiou, F. (2004). A phenomenological study of the long-term cosmic ray modulation, 850–1958 AD. *Journal Geophysical Research, 109,* A12103. https://doi.org/10.1029/2004JA010685.

Meier, M. M., & Matthiä, D. (2014). A space weather index for the radiation field at aviation altitudes. *Journal of Space Weather and Space Climate, 4,* A13. https://doi.org/10.1051/swsc/2014010.

Muscheler, R., Joos, F., Beer, J., Müller, S. A., Vonmoos, M., & Snowball, I. (2007). Solar activity during the last 1000 yr inferred from radionuclide records. *Quaternary Science Reviews, 26*(1–2), 82–97. https://doi.org/10.1016/j.quascirev.2006.07.012.

NOAA. (2004). Halloween space weather storms of 2003, NOAA Technical Memorandum OAR SEC-88.

Parker, E. N. (1958). Dynamics of the interplanetary gas and magnetic fields. *The Astrophysical Journal Letters, 128,* 664. https://doi.org/10.1086/146579.

Parker, E. N. (1969). Galactic effects of the cosmic-ray gas. *Space Science Reviews, 9*(5), 651–712. https://doi.org/10.1007/BF00174032.

Posner, A. (2007). Up to 1-hour forecasting of radiation hazards from solar energetic ion events with relativistic electrons. *Space Weather, 5,* S05001. https://doi.org/10.1029/2006SW000268.

Reames, D. V. (1999). Particle acceleration at the Sun and in the heliosphere. *Space Science Reviews, 90,* 413. https://doi.org/10.1023/A:1005105831781.

Ritter, B., Meskers, A. J., Miles, O., Rußwurm, M., Scully, S., Roldán, A., et al. (2015). A Space weather information service based upon remote and in-situ measurements of coronal mass ejections heading for Earth-A concept mission consisting of six spacecraft in a heliocentric orbit at 0.72 AU. *Journal of Space Weather and Space Climate*, *5*, A3. https://doi.org/10.1051/swsc/2015006.

Scholl, H., Marzari, F., & Tricarico, P. (2005). Dynamics of Mars trojans. *Icarus, 175*(2), 397–408. https://doi.org/10.1016/j.icarus.2005.01.018.

Schrijver, C. J., & Siscoe, G. L. (2009). Prologue to: Heliophysics: plasma physics of the local cosmos. In C. J. Schrijver, & G. L. Siscoe (Eds.). Cambridge University Press. ISBN-13: 978-1107403222

Schrijver, C. J., Kauristie, K., Aylward, A. D., Denardini, C. M., Gibson, S. E., Glover, A., et al. (2015). Understanding space weather to shield society: A global road map for 2015–2025 commissioned by COSPAR and ILWS. *Advances in Space Research*, *55*(12), 2745–2807. https://doi.org/10.1016/j.asr.2015.03.023

Schwadron, N. A., et al. (2014). Does the worsening galactic cosmic radiation environment observed by CRaTER preclude future manned deep space exploration? *Space Weather, 12,* 622–632. https://doi.org/10.1002/2014SW001084.

Schwadron, N. A., Rahmanifard, F., Wilson, J., Jordan, A. P., Spence, H. E., Joyce, C. J., et al. (2018). Update on the worsening particle radiation environment observed by CRaTER and implications for future human deep-space exploration. *Space Weather, 16,* 289–303. https://doi.org/10.1002/2017SW001803.

Sodnik, Z., Heese, C., Arapoglou, P. D., Schulz, K. J., Zayer, I., Daddato, R., et al. (2017). Deep-space Optical Communication System (DOCS) for ESA's space weather mission to lagrange orbit L5. In *2017 IEEE International Conference on Space Optical Systems and Applications (ICSOS)* (pp. 28–33). IEEE. https://doi.org/10.1109/icsos.2017.8357207

Steinhilber, F., & Beer, J. (2013). Prediction of solar activity for the next 500 years. *Journal of Geophysical Research: Space Physics, 118,* 1861–1867. https://doi.org/10.1002/jgra.50210.

Tammann, G. A., Loeffler, W., & Schroeder, A. (1994). The Galactic supernova rate. *The Astrophysical Journal Supplement Series, 92,* 487–493. https://doi.org/10.1086/192002.

Usoskin, I. G. (2017). A history of solar activity over millennia. *Living Reviews in Solar Physics, 14*(1), 3. https://doi.org/10.1007/s41116-017-0006-9.

Vanatta, M., Moraguez, M., & Miller, D. (2018). Integrative shielding: Reorganization and trade evaluation of ECLSS and propulsion systems for radiation mitigation on deep space missions. In *48th International Conference on Environmental Systems.*

Vourlidas, A. (2015). Mission to the Sun-Earth L5 lagrangian point: An optimal platform for space weather research. *Space Weather, 13,* 197–201. https://doi.org/10.1002/2015SW001173.

Chapter 5
Mars and Beyond: The Feasibility of Living in the Solar System

Chris Impey

Abstract Mars has occupied a distinctive place in the popular imagination for a century. Science fiction and the wishful thinking of Percival Lowell primed us to think of Mars as a living world, but the first landers and orbiters witnessed a frigid and arid desert landscape. Now the pendulum has swung back toward habitability, with evidence of the red planet's warmer and wetter history and hints of sub-surface aquifers. A Mars base with short-term occupants would be a first step toward eventual colonization. NASA is planning to send astronauts to Mars, subject to its limited and uncertain budget. Meanwhile, the private sector has ambitious plans for establishing a foothold on Mars, with the Mars One plans viewed more skeptically by experts than the plans of SpaceX. Mars colonists would face challenges caused by isolation, radiation, reduced gravity, and an unforgiving external environment. Over time, they would diverge culturally, psychologically, and genetically from the inhabitants of Earth.

5.1 The Lure of Mars

What happened to the Mars of our imagination? In mythology, the red planet was ominous and threatening. Babylonian astronomers took note of Mars, with its reddish color and its strange, episodic backward motion on the sky. They called it Nergal, god of the underworld and bringer of plague, epidemics, and disaster. The ancient Greeks associated Mars with Ares, one of the 12 Olympians and a son of Zeus and Hera. Ares was a violent and spiteful god who enjoyed combat; his sister Athene called him "a thing of rage, made of evil, a two-faced liar." The Greeks refused to honor him and no sacred places were built in his name. He was remembered on the battlefield, where his companions were Deimos and Phobos, the gods of terror and fear. His reputation softened slightly with the Romans, who made him the god of agriculture as well as war.

C. Impey (✉)
University of Arizona, Tucson, AZ, USA
e-mail: cimpey@as.arizona.edu

K. Szocik (ed.), *The Human Factor in a Mission to Mars*,
Space and Society, https://doi.org/10.1007/978-3-030-02059-0_5

The Dutch astronomer Christiaan Huygens drew the first map of Mars in 1659, and in a posthumous book called *Cosmotheoros* speculated that the bright spots on Mars were evidence of water and ice. He also thought that intelligent life could exist there. A century later, William Herschel demonstrated that Mars had seasons like Earth's and argued that ice at the poles could support life. In the mid-nineteenth century large telescopes produced sharper images of Mars and some astronomers thought that the dark colorations might be due to vegetation and the striations or streaks on the planet might be artificial constructions.

Percival Lowell had no doubts at all. The Boston merchant and keen amateur astronomer used his fortune to build a new telescope at a pristine, dark site in northern Arizona. He was racing to complete the telescope in time for a particularly close approach of Mars in 1894. Lowell convinced himself the features he saw on Mars were canals, where a dying civilization was trying to bring water from the poles to the equatorial regions. He wrote several sensational books to support his claims. A few years later, H.G. Wells used aspects of Lowell's work in his science fiction novel *War of the Worlds*. It was an instant classic. Once again Mars was malevolent: "…across the gulf of space, intellects vast and cool and unsympathetic, regarded our planet with envious eyes, and slowly and surely drew their plans against us" (Wells 1898).

In the early twentieth century, the scientific and pop culture views of Mars diverged. Edgar Rice Burroughs wrote "A Princess of Mars" as a serial in 1912 and as a book five years later. Civil War veteran John Carter mysteriously finds himself on Mars, a world populated with four-armed aliens, wild monsters, and scantily clad princesses. Carter uses weak gravity to exercise superhero powers and he gets the girl in the end. Burroughs wrote ten more Mars stories and his lurid fantasies inspired Arthur C. Clark and Ray Bradbury to launch a grand tradition of Mars science fiction. In 1938, Orson Welles revisited *War of the Worlds* with a radio show. His realistic live broadcast scared tens of thousands of people in the greater New York area; many left their homes at the prospect of a Martian invasion.

Meanwhile, Alfred Russel Wallace, co-discoverer of natural selection, had rebutted Percival Lowell that a freezing Mars could never support liquid water. This argument got stronger with remote sensing in the middle of the century. Mars fever finally cooled in 1965, when Mariner 4 swooped within 10,000 km of the planet's surface and saw an arid, crater-pocked terrain with no signs of life. The twin Viking landers cemented this picture in 1976.

Since then, the pendulum has swung back towards the middle. Mars can't host liquid water on the surface; a cup of water on the surface would evaporate in seconds because the typical temperature is −60 °C. It would evaporate rather than freeze because the atmosphere is so thin, very close to vacuum. The topsoil can't host life because the thin atmosphere allows it to be blasted by micrometeorites and sterilized by UV radiation and cosmic rays. But there's abundant evidence from orbiters of erosion and river deltas and shallow seas in the past. Craters overlying these features suggest that three billion years ago Mars was warmer and wetter and had a thicker atmosphere. A series of intrepid rovers—the Tonka toy size Sojourner in 1997, the go-kart sized twins Spirit and Opportunity starting in 1993, and the SUV-sized Curiosity

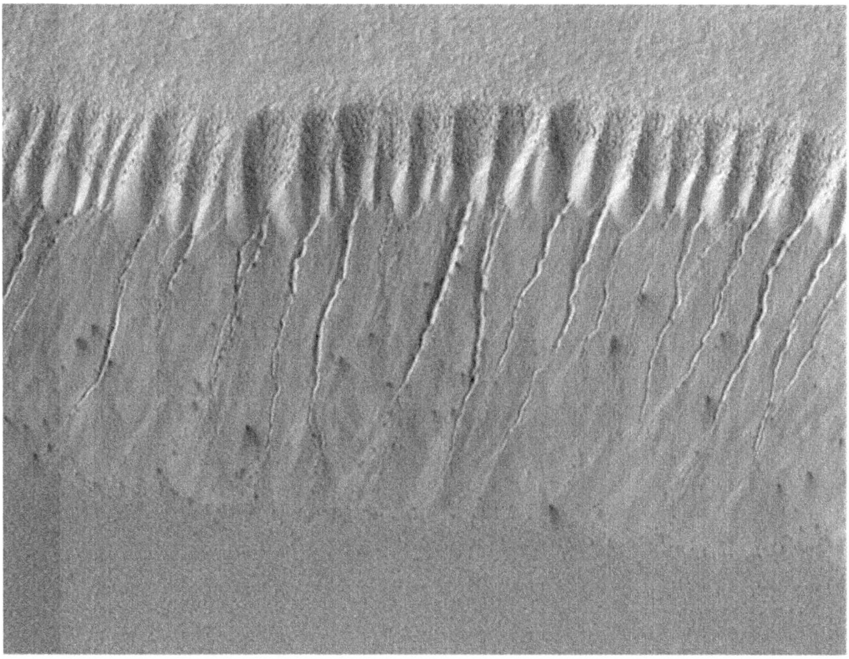

Fig. 5.1 The sharp v-shaped gullies on this escarpment on Mars are strong evidence of liquid water runoff. The water is likely to have seeped out of the escarpment a third the way down from the top, where it can be kept liquid by pressure in aquifers (NASA)

currently—have painted this picture in more detail. These rovers have collected rock samples that can only have formed in the presence of water (Grotzinger 2013; Hecht 2002).

Astronauts on Mars wouldn't find water on the surface but they'd be able to gather all they needed by a careful choice of the landing site. Spectroscopy from orbiting spacecraft reveals extensive ice at high latitudes, hidden by a veneer of dust and rock. If melted, and there was a protective atmosphere, it would cover the planet with a puddle deep enough to get your ankles wet. There's also evidence of gullies, with channels carved by water that erupts episodically from aquifers under the surface, where water can be kept liquid by pressure from above and radioactive heating from below (Fig. 5.1). There's active debate over how deep colonists would have to drill to hit water, but it might be as little as 10 or 20 m.

While it's not the Mars of anthropomorphized aliens or comic book superheroes, it's a planet that could host microbial life right now, as well as in the past. The red planet is far more habitable than the Moon because of the atmosphere, the higher temperature, and the amount of sub-surface water (Carr 1996). It beckons us to visit, and perhaps stay.

5.2 Establishing a Colony

Robert Zubrin never lost the faith. With a Ph.D. in Nuclear Engineering and over 200 technical papers to his credit, Zubrin has been a staunch advocate of human exploration of Mars for 30 years. He holds patents for hybrid rocket-planes, synthetic fuel manufacturing, magnetic sails, salt-water nuclear reactors, and three-person chess, but his true passion is Mars. He thinks we can lower the cost and complexity of a Mars mission by "living off the land," or utilizing many resources as possible from the air and soil. His ideas were strong enough to be adopted by NASA as their "design reference mission," but he became frustrated at NASA's glacial progress and anemic government support so he founded the advocacy group *Mars Society* in 1998. He's written a series of books that make the case for going to Mars (Zubrin and Wagner 1996; Zubrin 2008). His most recent book brings Mars exploration up to date with the Mars Direct proposal using the DragonX rocket (Zubrin 2013).

Asked about saving costs with a one-way journey, Zubrin has said: "Life is a one-way trip, and one way to spend it is by going to Mars and starting a new branch of human civilization there" (Zubrin 2011).

Mars is a challenging goal for human exploration. The problem isn't energy. The energy cost of going to Mars is less than 10% more than the energy cost of going to the Moon. The problem is the distance. An energy-efficient trajectory involves a travel time of 9 months each way. The trip can be shortened to 6–7 months at the expense of extra energy—a far cry from the week it takes to get to the Moon. The cost of transporting 2 years of supplies for even a small crew is daunting. Wernher von Braun was the first to make a technical study of a Mars mission in the 1950s but it was hopelessly grandiose, using a thousand Saturn V rockets to build a fleet of ten spacecraft in Earth orbit to then carry seventy astronauts to Mars. He pitched a scaled-down concept to Richard Nixon but it was passed over in favor of the Space Shuttle. Former NASA administrator Thomas Paine tried next. Perhaps he'd watched too much Star Trek, but he aimed to conquer and industrialize the Moon with nuclear space tugs, launch a fleet of space stations into orbit around the Earth, and send several dozen spaceships a year to Mars to build a space station and support the settlement The Reagan administration was happy to shelve his report.

In 2014, the National Research Council revisited human flight, as directed by Congress. Its sweeping 286-page report concluded bluntly that NASA had an unsustainable and unsafe strategy that will prevent the United States from achieving a human landing on Mars any time in the foreseeable future (National Research Council 2014). With current budgets, they suggest that it can't happen before mid-century. Along the way, the report addresses the philosophical question of why we should send people into space at all, concluding that purely practical and economic benefits don't justify the cost, but the aspirational aspect of the endeavor might make it worthwhile.

There must be good reasons and a strong will, because Mars is hard. One risk is radiation. Earth dwellers are sheltered from high-energy cosmic rays and solar flares by our atmosphere and magnetic field. When the Curiosity rover headed to Mars,

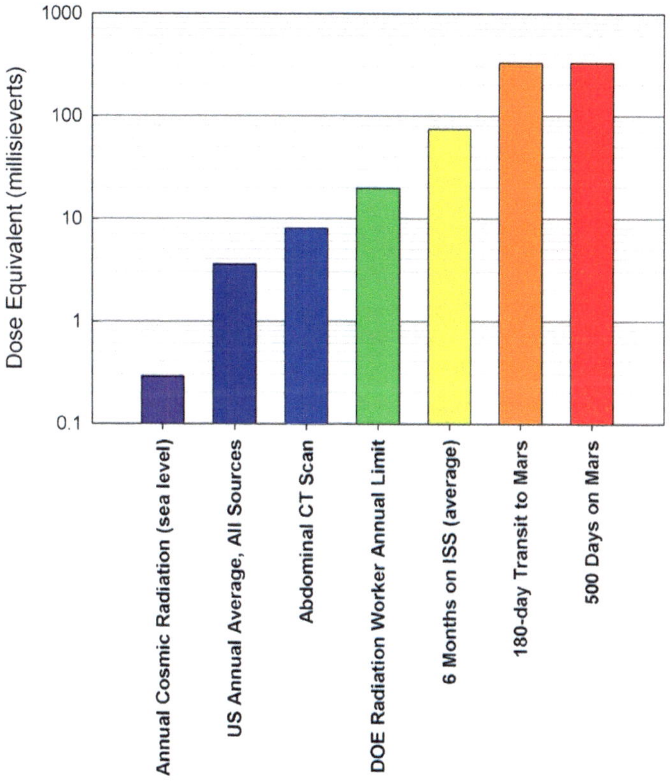

Fig. 5.2 Comparison of the exposure to high-energy cosmic rays in different situations; note the logarithmic scale. Most of the exposure comes from the 1-year transit time, equivalent to a hundred years of normal terrestrial exposure (NASA/JPL/Caltech)

scientists switched on a radiation detector and found that the radiation environment in deep space is far more intense than it is on Earth. An astronaut on a 2-year trip to Mars would get 200 times more radiation dose than an Earth dweller over that same period (Fig. 5.2). However, to put it in perspective, the adventure only increases the lifetime risk of cancer from 21 to 24%. The risk of some sort of spacecraft malfunction is likely to be much higher.

Another risk is weightlessness. Substantial physiological changes result from a microgravity environment. Russian cosmonaut Valeri Polyakov spent 438 days on board Mir, making a dizzying 7000 orbits of the Earth, in part to see if humans could handle a trip to Mars. The Russians reported that he suffered no long-term ill-effects from his 14 months in space. There is extensive literature on the adverse effects of microgravity on humans, including bone loss, muscle atrophy, cardiovascular dysfunction, and reduced functioning of the immune system (White and Averner 2001). Some of these effects, like bone loss, can be mitigated but not completely compensated for, by exercise and diet (Grimm et al. 2016).

Robert Zubrin notes that the used upper stage of a Mars launch vehicle could be employed as a counterweight. With a mile-long tether and a spin rate of 2 rpm, Earth gravity would be simulated. With a spin rate of 1 rpm, it would be Mars gravity and the astronauts could get acclimatized to the new situation before landing. Materials exist with the requisite tensile strength to construct such a tether, but it would add cost to a mission so it is not clear if such technology is warranted by the health risks.

A third risk is being cooped up. A Mars traveler would have to spend a year and a half in a cabin the size of a school bus, and as much as a year at their destination in a space no bigger than a large motor home. The Mars500 mission locked an international crew of six volunteers in a mock spaceship bound for Mars, but actually sitting in Moscow for a year and a half. The crew "returned to Earth" in 2011. Most of them experienced severely disrupted sleep patterns and all of them reduced their activity levels in the confined space, something researchers call a behavioral torpor (Vigo et al. 2013). The experiment made clear how important it will be to simulate Earth life rhythms in the spaceship or on Mars, and how important it will be to stay physically active.

It's hard to judge the psychological impacts of such a trip. People who winter in Antarctica experience a diluted version of the problems. But travelers to Mars will be the most isolated humans who ever lived. They'll have real-time interactions with a small number of companions and delayed communications with friends and loved ones who are tens of millions of miles away. They'll be in a confined space with no option to simply go out for a walk, and they'll be continuously monitored by anxious ground crews and scientists on Earth. If anyone spins out of control, there's no real-time access to mental health services such as counseling or psychotherapy.

The visionaries are undeterred. Apollo astronaut Buzz Aldrin put it like this: "Going to Mars means staying on Mars—a mission by which we are building up a confidence level to become a two-planet species. At Mars, we've been given a wonderful set of moons which can act as offshore worlds from which crews can robotically preposition hardware and establish radiation shielding on the Martian surface to begin sustaining increasing numbers of people" (Aldrin 2013).

Two new ventures are trying to put Mars within reach without using any government resources. Inspiration Mars is the brainchild of Dennis Tito, an engineer turned tycoon who was the world's first space tourist in 2001. Tito plans to keep costs down by not landing. His billion-dollar fly-by plans to use an upgraded version of the SpaceX Dragon capsule. With a cleverly designed trajectory, he can get there with a single burn of the engine. The return is challenging. The capsule will slam into the Earth's atmosphere at 32,000 mph, requiring new materials for a heat shield. The project is currently aiming for a launch in 2021.

Mars One is run by Dutch entrepreneur Bas Lansdorp, who also plans to use a SpaceX capsule. He plans to keep costs down by leaving his four passengers on Mars. If they survive the trip, they will build a habitat from their spacecraft and adjacent inflated areas covered by Martian regolith. They'll create water, oxygen, and some food locally, augmented by regular supply missions, and every 2 years they will be joined by four more refugees from Earth. Gradually, they will build a settlement (Fig. 5.3). Lansdorp estimates his costs to be $6 billion for the first trip and $4 billion

Fig. 5.3 Mars One is a Netherlands-based non-profit organization with plans to establish a permanent Mars settlement or a colony by 2024. The one-way trip would the subject of a reality TV series. (NASA Glenn Space Center)

for each crew that follows. Space experts judge the plan to be very ambitious; some judge it to be impossible. Everyone agrees that it is audacious (Do et al. 2014). NASA has a plan that will take several decades and cost about $100 billion, which makes the claims of Mars One seem unrealistic.

Would-be Martians are in a race against time. The red planet has its next close approach to the Earth in 2018, and it won't get as close again until 2035. Inspiration Mars and Mars One have both had to slip past the most favorable 2018 launch date. Mars One accepted over 200,000 applications online for the chance to live and die on Mars. In 2014 the number was culled to 1058, and then to 705. Those who remain will go endure rigorous physical and psychological testing to generate a final group of 24. Lansdorp plans to finance his venture by turning it into a reality TV epic—think *Survivor* meets *The Truman Show* meets *The Martian Chronicles*.

5.3 Greening the Red Planet

Let's ignore for a moment the evil twin. Venus is closest to the Earth in size and mass, and it has the same inventory of carbon dioxide. But on our planet most of the carbon dioxide is built into rocks and dissolved in the oceans, making them mildly acidic, leaving a moderately thick atmosphere to smooth out daily and seasonal temperature variations.

On Venus, only 30% closer to the Sun, the carbon dioxide built up in the atmosphere, triggering a runaway Greenhouse effect and raising the surface temperature to a level where lead melts. Whoever named Venus after the goddess of love had a sad history of relationships.

Mars is the misbegotten sibling, the runt. It's half the size of the Earth with a third the gravity. The next nearest Earth-like planet is tens of trillions of miles away, and unreachable with any current technology. The siblings went on divergent paths. One rusted and turned red, the other got the spark of life and turned green. Mars suffocated and dried out as its water and air leached into space, and it became scoured by dust storms and cosmic rays. Yet, it's at the edge of habitability, not much more inhospitable than a volcanic vent or a plateau in the high Andes. Mars has sunlight and reservoirs of water, carbon, nitrogen, and oxygen. One planet lived and the other died.

Perhaps we can make it live again? One of the most audacious ideas in science is planetary engineering. Planets don't stay the same. Geological evolution, combined with the aging of their parent star, can render a wasteland habitable and an Eden uninhabitable. This evolution occurs on geological timescales of hundreds of millions or billions of years.

Here's how the Earth has changed. It formed 4.5 billion years ago and minerals show that there was liquid water within 100 million years, so conceivably life started then. If it did, it must have survived the "Late Heavy Bombardment" 3.9 billion years ago, when unstable orbits in the Solar System led to surge of meteor impacts. Life around that time was limited to prokaryotes, or cells without nuclei, and there was no oxygen in the atmosphere. Around 3 billion years ago bacteria evolved that produce oxygen as a waste product, which is poisonous to other kinds of bacteria. The oxygen content of the atmosphere rose and 1.9 billion years ago eukaryotes evolve, or cells with nuclei. Life diversifies as it becomes multicellular and begins to reproduce sexually. Dramatic episodes of glaciation almost obliterate life 2.7 billion and 700 million years ago. In the last 10% of the chronology, life finally becomes big enough to see without a microscope, plants and animals evolve, they move onto the land, and a crescendo of evolution leads to mammals, primates, and finally us. The gist of this story is that dramatic change is normal for a biological world (Knoll 2004; Bjornerud 2005).

More recently, we have been inadvertently altering our own planet through industrial growth and the use of fossil fuel. Terraforming is the process by which we might potentially alter a different planet to make it more Earth-like or habitable by terrestrial life forms.

The first step would be to raise the temperature on Mars just enough to release frozen carbon dioxide from the polar regions, triggering a runaway Greenhouse effect. The positive feedback of this effect favors terraforming. While the carbon dioxide atmosphere of Mars only has 1% of the pressure of the Earth's atmosphere at sea level, there is enough carbon dioxide frozen in the soil to raise the pressure to 30% of the Earth's. Robert Zubrin and Chris McKay have outlined several ways to accomplish this. One is to fabricate a 100-kilometer mirror to direct extra sunlight toward the poles. Even if made from aluminized Mylar such a mirror would weigh 200,000 tons. Being too heavy to launch from Earth, this would have to be constructed from materials refined on Mars. Another method is to produce efficient heat-trapping gases on Mars, using industrial-scale facilities. There's a rich irony in using methods to make Mars habitable that are in danger of rendering the Earth uninhabitable. These two methods would each use as much energy as a city like Denver or Seattle, and need hundreds of workers to implement. A clever, less costly idea is to redirect small asteroids to impact the surface of Mars. Carbon dioxide would be liberated by heat energy from the impact, and asteroids can deliver ammonia, a very efficient greenhouse gas, and dust, which will cause Mars to absorb more sunlight (Zubrin and McKay 1993).

The next step is to activate a hydrosphere: raise the temperature by an additional amount sufficient to allow liquid water on the surface. Although still inhospitable, these conditions would allow extremophile microbes such as lichen, algae, and bacteria to be established. Their role is to prepare the regolith for photosynthetic organisms. Microbes used for this will be engineered to be optimally suited for their job. If the heating is done with asteroid impacts, these first two steps might take two to three hundred years.

The last step is to add oxygen to the atmosphere. Since oxygen is flammable, care would have to be taken to also add a buffer gas like nitrogen. Brute force would have to be used to import or create the initial oxygen needed for primitive plants, but once more advanced plants can propagate, they become the engine for oxygen production. It would take 500–1000 years to make an atmosphere suitable for animals or humans.

Terraforming may be possible and it's exciting at a technical level, but to see life breathed into the idea, we can turn to fiction. Kim Stanley Robinson wrote a trilogy of science fictions books in the mid-1990s that told the story of an overpopulated and dying Earth and the "First Hundred," a pioneering group of Mars colonists. The books capture the ethical issues we'll face if we go there, telling of the tensions between the Reds who prefer to leave Mars in its pristine state and the Greens who want to turn the planet into a second Earth. The trilogy covers colonization, terraforming, and the long-term future of human habitation on Mars (Robinson 1993, 1994, 1995).

The storytelling is very entertaining, but the physical descriptions are beyond evocative; they're mesmerizing. Who wouldn't want to visit Mars after reading this excerpt from *Red Mars*: "The sun touched the horizon, and the dune crests faded to shadow. The little button sun sank under the black line to the west. Now the sky was a maroon dome, the high clouds the pink of moss campion. Stars were popping out everywhere, and the maroon sky shifted to a vivid dark violet, an electric color that was picked up by the dune crests, so that it seemed crescents of liquid twilight lay across the black plain."

5.4 Extending Our Senses

What if we could have the experience of space travel without actually making the journey? The cost and difficulty of protecting fragile humans and sending them vast distances through space suggest we should find a different way to explore space. To see an alternative, look at the evolution of video games.

Pac-Man was the most famous arcade video game of all time. Released in 1980, the game had the player steer a small colored icon through a maze eating dots. *Pac-Man*'s popularity eclipsed that of space shooter games like *Space Invaders* and *Asteroids* and it's been estimated that by the end of the 20th century ten billion quarters had been dropped in *Pac-Man* slots. In 2000, a new computer game came out where the player could create virtual people, houses, and towns and watch their cartoon characters live their virtual lives. *The Sims* sold more than 150 million copies worldwide. If we think of how far video games came in 20 years, from the primitive graphics of *Pac-Man* to the cartoonish but quasi-realistic 3D graphics of *The Sims*, imagine what another 20 years will bring. A hint of that came in 2014 with the release of the Oculus Rift, a gaming helmet that immerses a player in 3D virtual reality. The best sense of the experience is the dramatic opening sequence of the 3D movie *Gravity*.

The future of Solar System exploration may lie in telepresence, a set of technologies that allow a person to feel that they're in a remote location. Video conferencing is one familiar and simple form of this technology. The market for projecting images and sound to connect meeting participants from around the world is growing 20% a year and is worth nearly $5 billion. Skype video calls now account for a third of all international calls, a staggering 200 billion minutes a year. Other examples include using robots with sonar to explore the ocean floor or robots with infrared sensors to explore caves. The robot provides the "eyes and ears" for an operator who doesn't have to leave the comfort of an office or their home.

When we "look" at Mars through the camera eye of the Curiosity rover or "sniff" the atmosphere with its spectrometer we are using a form of telepresence. NASA has used red-green stereoscopic imaging on all of its recent rovers but it missed a big chance to grab the public eye when it failed to build a 3D high-definition video camera into the Curiosity rover in time for launch. Film director James Cameron had pitched the camera to give Earthlings "you are there" immediacy as the rover trundled around the red planet. Intriguingly, telepresence doesn't have to convey the remote scene with perfect fidelity, because the brain has a tendency to "fill in the blanks" and "smooth out the rough edges" of any visual representation that is familiar (Rodriguez-Ardura and Martinez-Lopez 2014).

A lot has changed since the last Apollo astronaut walked on the Moon. At the time of the moon landing, real-time, complex decision-making had to be carried out by people. Now, robots and machines have impressive capabilities and they can be remotely controlled by scientists far away.

Planetary scientists have used remote sensing for 40 years. The twin Viking landers were designed to analyze samples of Mars soil for traces of microbial life. No cameras were included in the specification. But Carl Sagan argued that images from the surface would engage the public. Besides, he noted mischievously, what if there are Martian polar bears and we miss them because we don't take pictures? So they were added, and the images of stark desert vistas that they returned were immediately compelling to the public. Probes to the outer Solar System since then have "watched" the volcanoes on Io, "listened" to magnetic storms on Jupiter, "sniffed" the thick atmosphere of Titan, and "tasted" the icy geysers on Enceladus.

Telepresence implies something more than remote sensing; it's a technology that allows someone to feel like they're in a remote location. The word was coined in 1980 by U.S linguist and cognitive scientist Marvin Minsky. He was inspired by a short story by science fiction author Robert Heinlein. The concept was further developed by Fred Saberhagan in *Brother Assassin* from the Berserker series: "…it seemed to all his senses that he had been transported from the master down into the body of the slave-unit standing beneath it on the floor. As the control of its movements passed over to him, the slave started gradually to lean to one side, and he moved its foot to maintain balance as naturally as he moved his own. Tilting back his head, he could look up through the slave's eyes to see the master-unit, with himself inside, maintaining the same attitude on its complex suspension" (Saberhagen 1997).

This level of control and verisimilitude is far off in space exploration, but we're approaching it with the virtual reality of video games. The difference between gaming and science applications is that a video game tries to digitally recreate a real-world experience while science uses technology to digitally represent and transmit the real world.

Remote control of robots—often called telerobotics— is infiltrating life in surprising ways. Robots are used to defuse bombs, extract minerals from hazardous mines, and explore the deep sea floor. They also act as aerial drones and doctor's assistants. They're even beginning to be seen in the boardroom and the workplace. Many commercial robots look like vacuum cleaners with a screen on top and they are no more than ventriloquist's dummies, but after the comical first impression it's disconcerting to realize that there's a real person at the other end of the device. A striking recent example was a talk by Edward Snowden at the TED2014 conference (Snowden 2014). The controversial NSA whistleblower was in hiding somewhere in Russia, but he was represented on stage by a screen attached to two long legs that ended in a motorized cart. Snowden communicated with the moderator and turned toward the audience to answer questions; he could see and hear everything that was going on.

At a 2012 symposium on "Space Exploration via Telepresence" held at NASA's Goddard Space Flight Center, scientists rubbed shoulders with roboticists and technology entrepreneurs. A major topic was latency: the time it takes a robot to respond to commands and communicate results back to the operator. Latency is governed by the speed of light. In terrestrial applications, latency is essentially zero, but on the Moon it's a couple of seconds, on Mars it ranges from 10 to 40 min, and to the outer Solar System it's up to 10 h. This makes real-time communications impossible.

Astronauts on the International Space Station have tested the remote control of a mobile robot called Justin, which was developed by the German Aerospace Center (Dietrich et al. 2013). The robot has four-fingered hands and astronauts control it with a sense of "touch." This is done with haptic technology that uses fields, vibrations, and forces to recreate the feeling of touch (Grunwald 2008). To avoid latency, and to avoid the costs of going in and out of gravity pits, future explorers may control ground operations from Moon orbit or Mars orbit. NASA is testing a "blue collar" robotic miner that digs, fills and empties buckets, and can right itself if it falls. It would be part of the advance expedition to Mars to mine and build using local materials

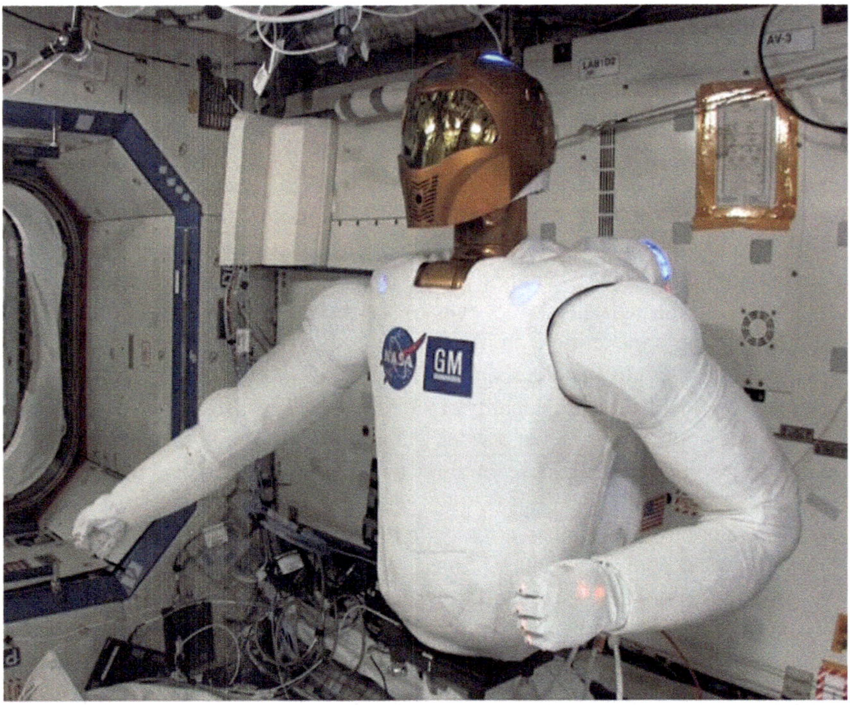

Fig. 5.4 Robonaut 2 (also known as Justin) works on the International Space Station in 2011, controlled by astronaut Mike Fossum. Routine tasks in space will increasingly be taken over by programmable and remotely controllable robots (NASA)

in preparation for the later arrival of astronauts. Meanwhile, the European Space Agency is developing a robotic exoskeleton so that astronauts can control a remote robot as if it was an extension of their body, and they've already tested a robot that can carry out simple tasks in the International Space Station (Fig. 5.4).

The frontier of telepresence is its merger with artificial intelligence, a development foreseen by computer science pioneer Marvin Minsky nearly 40 years ago (Minsky 1980). A robot doesn't need to just be a remote extension of a human; it can process information and makes its own decisions. This will be exciting but it will raise fascinating moral and ethical questions, especially if these semiautonomous robots come into contact with each other.

To some, telepresence is a retreat from the grand vision of personal experience and direct human exploration. For others, telepresence is the way we can enhance our capabilities and extend our senses while dealing with the harsh environments that lie beyond the Earth (David 2013).

5.5 Space Societies

Experts agree that it will be far easier and cheaper to fix the problems of this planet than to find a way to live off-Earth. What are the challenges that might want to make us find a new home in space? The ultimate demise of Earth will occur in four billion years when the Sun runs out of its nuclear fuel. At that point, the Sun's core will collapse and the star's violent reconfiguration will eject a layer of gas that will engulf the Earth and cook the biosphere. But long before that, the Sun will start to burn hotter as it consumes its hydrogen and about a half billion years from now the temperature on Earth will have risen enough to make the oceans boil (Schroder and Smith 2008).

The grim fate of the biosphere is a reminder that habitable zones evolve on long timescales. A planet is kept habitable by a symbiotic relationship between living organisms and the rocks and oceans, an insight first noted by James Lovelock, who originated the Gaia hypothesis. The fundamental drive of habitability is energy from the parent star. The Sun was 25% dimmer 3–3.5 billion years ago, when life on Earth was microbial and oxygenic photosynthesis has not yet evolved. In the future, as the Sun uses it nuclear fuel and settles into a more compact configuration, it will "burn hotter" so the Earth will not remain habitable for the full span the Sun has nuclear fuel, another 4–4.5 billion years. Microbes that live inside the rock or far below the surface of the ocean are immune from moderate changes in solar radiation since they live off geological energy. As the Earth becomes intolerably hot, we'll have to develop Biospheres for the whole population, assuming humans persist that long.

Those timescales are long enough that we might be forgiven for not getting too worried. The best metric for proximate danger is the Bulletin of the Atomic Scientists. Starting in 1947, a group of scientist and engineers created a Doomsday Clock to show how far we were from apocalypse. As the threat of a nuclear catastrophe receded, the proximity of the clock to midnight started to take into account the possibility that through climate change, biotechnology, or cyber-technology we could cause irrevocable harm to our way of life and the planet. The clock sat at 2 min to midnight in 1953, at the nadir of the Cold War. It receded to 17 min from midnight with the fall of the Soviet Union. But now it reads 5 min to midnight because of a surge of nuclear weapons in the hands of small, unstable countries, and the sense that climate change may have passed a tipping point.

Since 2012, the Doomsday Clock has stood at 5 min to midnight, uncomfortably close to disaster. The explanation associated with that judgment is worth quoting in full:

> The challenges to rid the world of nuclear weapons, harness nuclear power, and meet the nearly inexorable climate disruptions from global warming are complex and interconnected. In the face of such complex problems, it is difficult to see where the capacity lies to address these challenges. Political processes seem wholly inadequate; the potential for nuclear weapons use in regional conflicts in the Middle East, Northeast Asia, and South Asia are alarming; safer nuclear reactor designs need to be developed and built, and more stringent oversight, training, and attention are needed to prevent future disasters; the pace of

technological solutions to address climate change may not be adequate to meet the hardships that large-scale disruption of the climate portends (Bulletin of the Atomic Scientists 2018).

Many voices have weighed in on leaving the Earth. In 1895, the rocket pioneer Konstantin Tsiolkovsky said “Earth is the cradle of humankind, but one cannot live in the cradle forever.” Carl Sagan put in this way a century later: “Since, in the long run, every planetary civilization will be endangered by impacts from space, every surviving civilization is obliged to become spacefaring—not from exploratory or romantic zeal, but for the most practical reason imaginable: staying alive.” Science fiction writer Larry Niven was more succinct, “The dinosaurs became extinct because they didn’t have a space program.” We may be able to fend off impacts from space but physicist Stephen Hawking sounds the alarm about other threats: “It will be difficult enough to avoid disaster in the next hundred years, let alone the next thousand or million. Our only chance of long-term survival is not to remain inward-looking on planet Earth, but to spread out into space” (Tsiolkovsky 1911; Sagan 1994; Clarke 2001; Hawking 2012).

A mass exodus from Earth is implausible. After all, it costs $50 billion to send a dozen people to the Moon for just a few days. Elon Musk may claim he’ll reduce the price of a trip to Mars to $500,000, which is a hundred thousand times less, but that seems an unlikely extrapolation of progress at the moment. If the Earth becomes contaminated or inhospitable we’ll have to live in bubble domes, fix it, or suffer through it. But in this century, a first cohort of adventurous humans will probably cut the umbilical and live off-Earth. What issues will they face?

Beyond survival, their first issue is their legal status. The Outer Space Treaty of 1967 addresses ownership. According to Article II, “Outer space, including the Moon and other celestial bodies, is not subject to national appropriation by claim of sovereignty, by means of use or occupation, or by any other means.” That seems transparent, but it doesn’t mention the rights of individuals. Bas Lansdorp, the CEO of Mars One, said his legal experts looked into the treaty and he thinks that, “What goes for governments also goes for individuals in those governments.” If Mars One achieves its goal, 30 people will settle in the red planet by 2023; the gradually expanding settlement will use more and more Martian land. Realistically, a Mars settlement will not be built until well into the 2030s. Lansdorp insists that their goal isn’t ownership. “It is allowed to use land, just not to say that you own it,” he says. “It is also allowed to use resources that you need for your mission. Don’t forget that a lot of these rules were made long ago, when a human mission to Mars was not within reach” (Wills 2013).

Some space players claim altruistic motives, but none of them can succeed without revenue to fuel their dreams. What happens when profit is the only goal? A large, multinational corporation is bound by international trade law but they could plausibly argue that they have the right to use, even to exhaust, the resources of an extraterrestrial body. A government that wanted to appropriate land on the Moon or Mars might withdraw from the Outer Space Treaty, and it’s unlikely they would suffer any serious consequence. Even Mars One exists in a legal limbo. Bas Lansdorp needs to fund his $6 billion mission: "Imagine how many people would be interested in a grain of sand from the New World!"

At some point, the debate will stop being hypothetical. The history of colonization of the Earth shows that a claim of ownership is irresistible. Each succeeding generation of settlers who are born and die beyond Earth will feel less connection to the home planet. They may chafe at the rules and regulations imposed from afar. Tanja Masson-Zwaan is Deputy Director of the International Institute of Air and Space Law and a legal advisor to Mars One. She says, "I assume at some point these settlers will become more detached from Earth, and will live by their own rules."

The historical example of manifest destiny is misleading in the context of space colonization. Countries have grown and gained resources on Earth by seizing territory and displacing or subjugating the original inhabitants. Even in the twenty-first century, the stains of this brutal history persist. Space is a new resource. The people who leave Earth won't be taking land from anyone (Globus 1975). Eventually, they'll have to make everything they need to survive and prosper. They will create their own wealth. It will be hard to hold them to any Earth-centric legal framework if they want to be independent.

Colonization implies replacement and growth. A Mars colony can be augmented by new arrivals but a healthy, normal culture centers on the family unit. There will be sex and there will be babies.

Sex in space hasn't progressed beyond snickering and titillation. It's the stuff of urban, orbital legend. Every couple of years, NASA and its Russian counterpart wearily deny that astronauts have had sex. The astronauts themselves stay tight-lipped. Official policy forbids it. Zero gravity sex is tricky for several reasons. Blood flow doesn't work as well as on Earth so men will have trouble getting erections. Sweat piles up in layers making intimacy less pleasant. Physics is also an obstacle; the slightest push or push sets an object in motion. NASA astronaut Karen Nyberg once demonstrated this by using a single strand of her hair to propel herself across the cabin. Straps and harnesses would have to be used, and given human ingenuity and desire, it's possible that intercourse has taken place in some quiet, dark corner of the International Space Station. But it's not written anywhere in the mission log.

Martian sex presents fewer obstacles. The 37% gravity would require minor adjustments. To finesse the issue of procreation, if not coupling, all-male or all-female crews have been proposed. More controversially, voluntary sterilization has been suggested for the first long-term visitors. Mars One plans to arms its visitors with contraceptives but it's not known how well they would work on Mars. Norbert Kraft, the medical director of the project, isn't entirely reassuring when he says they will "make colonists aware of the risks associated with having sex." The first waves of Mars visitors will die there and they know that the medical facilities will be rudimentary; they're unlikely to want babies. But as colonies get established the dictates of biology and human culture will prevail.

Even if we discount Mars One's plans as fantastical and grandiose, colonization is likely eventually because there are enough pioneers with financial backing to make it happen. When a small group of humans branch out from the root of the tree, who will they become?

5.6 Evolutionary Divergence

Imagine when the first baby is born off-Earth. That event will be an extraordinary milestone, resetting the clock of human existence. In Arthur C. Clarke's short story "Out of the Cradle, Endlessly Orbiting," an engineer at a Moon base is preparing to relocate to Mars when his wife goes into labor. The baby's first, plangent cry shakes him to his core, resonating more than the roar of any rocket ship (Clarke 1962).

How many people does it take to start over? In conservation biology and ecology there's a term called "minimum viable population." This is the lower bound on the population of a species in the wild such that it can survive natural disasters and demographic and genetic variations. In animal population studies, about 500 adults are required to avoid inbreeding, and 5000 adults are required to allow a species to pursue a typical evolutionary lifespan from origination to extinction of one to ten million years (Thomas 1990). These are rough estimates, used in biology to estimate the probability of extinction; in the United States similar models of minimum viable population trigger protection by the *Endangered Species Act*.

For humans, the minimum size can be relevant during a dramatic population bottleneck. If a species population is reduced by environmental catastrophe, the genetic diversity in the remaining individuals is less and it can only grow slowly by random mutations. The robustness of the remaining population is reduced, making them more vulnerable to another adverse event (Catton 2009). This is true even though the survivors may have been the fittest individuals. Also, inbreeding is more likely, with offspring having an increased chance of recessive or deleterious traits (Ramel 1998).

When geneticists sequenced the DNA of chimps and humans, they made the staggering discovery that a single band of 30–80 chimps can have more genetic diversity than all seven billion humans alive today (Dawkins 2004). We have very little genetic diversity even though it could have developed since we diverged from chimps six million years ago. Research on mankind's restricted gene variation indicates that humans migrated out of Africa about 88,000 years ago and at some stage before that our numbers may have dwindled to as low as two thousand. Some geneticists hypothesize that this bottleneck was caused by the explosion of the Toba supervolcano in Indonesia and resulting major environmental change (Templeton 2007). Other researchers think the bottleneck may have been more drawn out and not due to sudden environmental change, with numbers dropping as low as 2000 for tens of thousands of year until the Late Stone Age. Regardless of the cause, our genetic makeup hints to the fact that we were once in a perilous state, at the edge of extinction (Hawks et al. 2000; DeGiorgio et al. 2009).

More recent human history gives better examples of how to define the viable size of a space colony. When a new population is established by a small number of individuals from a larger population, it's subject to the founder effect, first described by evolutionary biologist Ernst Mayr. The founder effect leads both to loss of genetic variation and to genetic divergence from the original population.

In 1790, Fletcher Christian and 8 other mutineers from the H.M.S. Bounty were joined by 12 Polynesian women to settle on Pitcairn Island, a windswept volcanic island in the South Pacific. The 50 current residents are all descended from these few "founders." In 1814, 15 British voyagers settled the remote island of Tristan de Cuhna, located in the Atlantic midway between South Africa and South America. The population had grown to 300 by 1961, when a volcano erupted and everyone was evacuated to England. These small populations left the inhabitants subject to genetic abnormalities. Fletcher Christian spread a gene that contributes to Parkinson's disease on Pitcairn Island while the current inhabitants of Tristan de Cuhna have ten times the normal incidence of a degenerative eye that leads to blindness. You don't have to be stuck on an island or Mars to suffer genetic isolation. The 18,000 Old Order Amish of Lancaster, Pennsylvania are descended from a few dozen individuals who emigrate from Germany in the early 1700s. It's tragic that babies born into this community have a high incidence of an extremely rare and fatal genetic disorder called microcephaly (Macgregor et al. 2010).

The sweet spot for a space colony may be the size of a village. John Moore, an anthropologist from the University of Florida, has developed simulation software for analyzing the viability of small groups (Carrington 2002). He suggests the optimum number for a viable long-term colony is about 160. This number could probably be reduced with judicious genetic selection to minimize the probability of inbreeding.

If space colonists don't get "new blood" from the home planet, their gene pool will experience genetic drift. Genetic drift is the change in the frequency of gene variants or alleles due to random sampling. The effect is larger in smaller populations and it acts to reduce genetic variation, which in turn reduces a population's ability to respond to new selective pressures. This sounds bad, but genetic drift and the founder effect on Earth are major drivers of evolution. They lead to the formation of new species.

Over generations, the colonists will evolve. We can imagine some of the changes that will take place. The lower gravity on Mars will alter the cardiovascular system and reduce the cross-sectional area of load-bearing bones and tendons. There will be accelerated trends seen in human evolution on Earth—toward being taller, and having less body hair, weaker muscles, and smaller teeth. The lack of a varied natural environment would probably lead to weaker immune systems. An additional challenge will be to maintain sensory stimulation as well as intellectual stimulation, to keep the brain sharp (Finney and Jones 1985).

A new species will have evolved if off-Earth humans can no longer mate and produce viable offspring with those who never left Earth. We know this will take a long time because a small group of people went on a one-way trip to the Americas about 14,000 years ago, and when they encountered Europeans 500 years ago, they were still the same species. Some groups in Australia and Papua New Guinea have been mostly isolated for 30,000 years and speciation didn't occur. But for colonists on the Moon or Mars, the process will be accelerated by the very different physical environment and the higher incidence of mutations due to cosmic rays.

Finally, after hundreds of thousands of years and thousands of generations, when the first off-Earth baby's cry is no more than an ancestral memory, the colonists will have come of age. They will no longer be us. Imagine the colonists live in total isolation and one day we encounter the ancestors of the people who left our planet. They'll speak their own language, have their own culture, and only resemble us partly. For each side, it will be like looking in an eerily distorted mirror.

Acknowledgements This article is adapted from material previously published in the book Beyond: Our Future in Space, by Chris Impey, in 2015, with permission from the publisher, W. W. Norton and Company.

References

Aldrin, E. E. (2013). http://buzzaldrin.com/what-nasa-has-wrong-about-sending-humans-to-mars/.

Bjornerud, M. (2005). *Reading the rocks: The autobiography of the Earth*. New York: Basic Books.

Bulletin of the Atomic Scientists. (2018). The Doomsday clock and its timeline can be seen online at http://thebulletin.org/timeline.

Carr, M. H. (1996). *Water on Mars*. Oxford: Oxford University Press.

Carrington, D. (2002). Magic number for space pioneers calculated. *New Scientist*. http://archive.is/Xa8I. A report on the work of John Moore.

Catton, W. R. (2009). *Bottleneck: Humanity's impending impasse*. Xlibris.

Clarke, A. C. (1962). *Tales of ten worlds*. New York: Harcourt Brace.

Clarke, A. C. (2001). Meeting of the minds: Buzz Aldrin visits Arthur C. Clarke. Reported by Andrew Chaikin on February 27, 2001, from https://www.space.com.

David, L. (2013). Remote telepresence: A new tool for space exploration? *Aerospace America 51*(2), 38–44.

Dawkins, R. (2004). *"The Grasshopper's tale", The Ancestor's tale: A Pilgrimage to the dawn of life*. Boston: Houghton Mifflin.

DeGiorgio, M., Jakobsson, M., & Rosenberg, N. A. (2009). Explaining worldwide patterns of human variation using a coalescent-based serial founder model of migration outward from Africa. In *Proceedings of the National Academies of Science U.S.A.* (Vol. 106, No. 38, pp. 16057–16062).

Do, S. et al. (2014). An independent assessment of the technical feasibility of the Mars one mission plan. http://hdl.handle.net/1721.1/90819.

Dietrich, A., Ott, C., & Albu-Schaffer, A. (2013). Multi-objective compliance control of redundant manipulators: hierarchy, control, and stability. In *Proceedings of the 2013 IEEE/RSJ International Conference on Intelligent Robots and Systems*, Tokyo, Japan, (pp. 3043–3050).

Finney, B. R., & Jones, E. M. (Eds.) (1985). *Interstellar migration and the human experience*, University of California Press, Berkeley, California.

Globus, A. (1975). Space Settlement Basics, NASA Ames Research Center web pages of space colonization. http://settlement.arc.nasa.gov/Basics/wwwwh.html.

Grimm, D., Grosse, J., Wehland, M., Mann, V., Reseland, J. E., Sundaresen, A., et al. (2016). The effects of microgravity on bone in humans. *Bone, 87,* 44–56.

Grotzinger, J. (2013). Introduction to special issue: Analysis of surface materials by the curiosity Mars rover. *Science, 341*, 1475.

Grunwald, M. (2008). *Human haptic perception*. Berlin: Birkhauser Verlag.

Hawking, S. (2012). Quote from a transcript of a video interview for *BigThink*. http://bigthink.com/videos/abandon-earth-or-face-extinction.

Hawks, J., Hunley, K., Lee, S. H., & Wolpoff, M. (2000). Population bottlenecks and Pleistocene human evolution. *Molecular Biology and Evolution, 17*(1), 2–22.

Hecht, M. H. (2002). Metastability of liquid water on Mars. *Icarus, 156,* 373–386.

Knoll, A. H. (2004). *Life on a young planet: The first three billion years of evolution on Earth*. Princeton, New Jersey: Princeton University Press.

Macgregor, S., et al. (2010). Legacy of the mutiny on the bounty: Founder effect and admixture on Norfolk Island. *European Journal of Human Genetics, 18*(1), 67–72.

Minsky, M. (1980). *Telepresence, Omni Magazine*. http://web.media.mit.edu/~minsky/papers/Telepresence.html.

National Research Council. (2014). *Pathways to exploration: Rationales and approaches for a U.S. program of human space exploration* by the Committee on Human Spaceflight, National Research Council, Washington, DC.

Ramel, C. (1998). Biodiversity and intraspecific genetic variation. *Pure and Applied Chemistry, 70*(11), 2079–2084.

Robinson, K. S. (1993). *Red Mars*. New York: Random House.

Robinson, K. S. (1994). *Green Mars*. New York: Random House.

Robinson, K. S. (1995). *Blue Mars*. New York: Random House.

Rodriguez-Ardura, I., & Martinez-Lopez, F. J. (2014). Another look at 'Being There' experiences in digital media: Exploring connections of telepresence with mental imagery. *Computers in Human Behavior, 30,* 508–518.

Saberhagen, F. (1997). *Brother assassin*. New York: Tor Books.

Sagan, C. (1994). *Pale blue dot: A vision of the human future in space* (p. 371). New York: Random House.

Schroder, K.-P., & Smith, R. C. (2008). Distant future of the Earth and Sun revisited. *Monthly Notices of the Royal Astronomical Society, 386,* 155–163.

Snowden, E. (2014). TED talk. http://www.ted.com/talks/edward_snowden_here_s_how_we_take_back_the_internet.

Templeton, A. R. (2007). Genetics and recent human evolution. *International Journal of Organic Evolution, 61*(7), 1507–1519.

Thomas, C. D. (1990). What do real population dynamics tell us about minimum viable population sizes? *Population Biology, 4*(3), 324–327.

Tsiolkovsky, K. (1911). Quote from a letter, online in Russian at http://www.rf.com.ua/article/388.

Vigo, D. E., et al. (2013). Circadian rhythm of autonomic cardiovascular control during Mars 500 simulated mission to Mars. *Aviation and Space Environmental Medicine, 84,* 1023–1038.

Wells, H. G. (1898). *War of the Worlds*. London: Bell. Chapter 1.

White, R. J., & Averner, M. (2001). Humans in space. *Nature, 409,* 1115–1118.

Wills, A. (2013). Lansdorp quoted in "Is Mars For Sale?" for *Mashable.com*. http://mashable.com/2013/04/09/mars-land-ownership-colonization/.

Zubrin, R. M., & McKay, C. P. (1993). Technological Requirements for Terraforming Mars. A Technical report for NASA's Ames Research Center. http://www.users.globalnet.co.uk/~mfogg/zubrin.htm.

Zubrin, R. M., & Wagner, R. (1996). *The case for Mars: The plan to settle the Red Planet and why we must*. New York: Simon and Schuster.

Zubrin, R. M. (2008). *How to live on Mars: A trust guidebook to surviving and thriving on the Red planet*. New York: Three Rivers Press.

Zubrin, R. M. (2011). National Public Radio *Science Friday* interview. http://www.npr.org/2011/07/01/137555244/is-settling-mars-inevitable-or-an-impossibility.

Zubrin, R. M. (2013). *Mars direct, space exploration, and the Red planet*. New York: Penguin.

Part II
Ethical, Social and Cultural Contexts of the Human Factor in a Mission to Mars

Chapter 6
Human Extremophiles: Mars as a Camera Obscura of the Extraterrestrial Scientific Culture

Klara Anna Capova

Abstract The start of the century marked the beginning of a new era of scientific goals and technological achievements in space technologies, together with the new visions of space exploration ambitions. The enterprising plans of space travels and creation of first human settlement on Mars are frequently presented as the next proverbial giant leap for humankind. The chapter focuses on private space programmes and considers the motivations for future exploration of Mars from a sociocultural perspective. Further, it discusses why Mars is of such importance to public imagination today. In presenting a socio-science fiction, and regarding science as the primary mode of operating of such endeavour, the chapter argues that the Martian colonists will need to live in the extreme environments, survive unprecedented scenarios and in a way become organisms that thrive in extreme conditions, extremophiles, themselves.

6.1 To Go or Not To Go?

To go to Mars or not to go to Mars, that is the new Hamletian dilemma of the Solar System exploration. Yet, recently it rather appears, the question we should ask is when and who will set off to the Red planet and raise the flag on Mars first. We can expect an international space mission or private aerospace company to launch their spacecraft to Earth's neighbour. Especially, the non-governmental commercial space programmes introduce brave plans of space exploration, colonisation of the Solar System and the visions of multiplanetary futures as the next proverbial giant leap for humankind.

One of the much publically exposed and criticised private space endeavours is the Mars One Mission. Mars One foundation's vision is to establish a permanent human settlement on Mars, giving the participants a one-way ticket and to make their dream about living on Mars come true. According to the recent company status, funding is being raised to move the mission forward. With the vision of establishing

K. A. Capova (✉)
Department of Anthropology, Durham University, Durham, CA, UK
e-mail: k.a.capova@durham.ac.uk

K. Szocik (ed.), *The Human Factor in a Mission to Mars*,
Space and Society, https://doi.org/10.1007/978-3-030-02059-0_6

a permanent settlement on the Red Planet, Mars One's roadmaps say the first settlers will set off in 2031, never to return to their home planet. The second crew is planned to follow 2 years later. Additional crews and cargo deliveries are envisioned to follow the first two waves of explorers every 26 months (Mars One 2018a).

Elon Musk, popular American technopreneur, announced his vision of making humans a multiplanetary species just 2 years ago (Musk 2017). In 2002, Musk founded SpaceX, a private American aerospace manufacturer and space transportation services company. SpaceX has become the first private company capable of returning a spacecraft from low Earth orbit. Other SpaceX services include commercial satellite launches and the US government missions, such as NASA resupply to the International Space Station (ISS). The announcement of the SpaceX mission plans and Musk's visions of the human future are seen on the front pages of news servers worldwide.

Despite overly optimistic predictions, enthusiastic media coverage and the overall ignorance of possible setbacks, including the crew's safety and technical feasibility, the first question stands—why is there so much interest in Mars? Why so much advocacy for going there? Answers could be sought in understanding various motivations for colonising Mars. What Jacques Arnould calls 'the explorer's complex' (Arnould 2014) seems to be one of the prime driving forces and a strong selling point for the many media stories. At the same time, the new 'cultural narratives of spaceflight' (Billings 2007) are produced, recently in context with commercial space programmes. Mass media have repeatedly reported Stephen Hawking's' warning that humans need to colonise other planets as a necessity to escape the inevitable apocalypse (Knapton 2017). One year later, Musk presented the vision for his space programme as a must: we must colonise Mars to preserve our species in a third world war (Musk in 2018). There are other less apocalyptic means of exoplanetary marketing and private space exploration motivations. Those include the acquisition of resources and space tourism as the next steps of human expansion beyond Earth's orbit.

Parallel to the pro-colonisation, debate has emerged numerous concerns regarding the ethics of exploration, ownership, sustainable use of space, planetary protection, social responsibility and environmentalism. Valentine pointed to the existence of the exit strategy in the space capitalism (Valentine 2012). Fairén discussed the biological contamination to protect science (Fairén et al. 2017), Billings suggested Mars should not be colonised (Billings 2017). Milligan examined the argument that we should recognise private property rights duty to extend human life as not a necessary imperative (Milligan 2011). Other voices have been raised against selling space colonising and immortality (Slobodian 2015) or describing the dreams of space colonisation as irrational (Williams 2010). Persson discussed the economic liberty in space which should not include the right for corporations to rule over space settlements (Persson 2015). In raising the case against Mars colonisation, Koepsell warned not enough attention is paid to the safety and health of the crew and settlers. Specifically, regarding the manned one-way mission, there is a lack of control and ethical considerations (Koepsell 2017). The actual challenge remains of how to ensure the survival of human explorers during the interplanetary flight and later on Martian surface, to be specific, to shield them from radiation. A recent study shows that exposure to space radiation may significantly damage gastrointestinal function in astronauts (Kumar et al. 2018). Since life could have existed on Mars, and we still seek for more compelling evidence, the preservation of the pristine environment of Mars and the protection of science interests become of importance to astrobiology research (Capova et al. 2018).

This chapter begins with a question why is there so much interest in Mars? Assuming that the dominant motivation is the explorer's complex, then the accounts of expeditions, the historical and cultural parallels in thinking about Mars and in general in envisioning the extraterrestrial futures can provide useful lessons. The New World is the first thing that comes to mind when thinking about legendary expeditions of the past. Notably that of Christopher Columbus, who in search of a route to India crossed the Atlantic to discover the new continent in 1492, where we date the start of the systematic European colonisation of the Americas. Recent developments in space commerce inspire thinking about the politics of space exploration, the emergence of the space gold rush. The outer space objects are regarded as exploitable natural resources (Olson 2012) and the new space age that one of outer space economy and private interests has emerged (Capova 2016).

In the case of Mars, aside of human expansionism, why is the Red Planet so appealing to human sentiments? Even the terrestrial analogue, the most unforgiving environment on Earth, Antarctica (and Arctic regions), is in comparison to Mars simply not adverse enough. To be able to survive on Mars, where there is no air to breathe, no water to drink and no food to eat, the future explorers will be left with what they will bring or produce. Significant investments and efforts are required to transport human life to the other planet. Why, when we know the robotic exploration of the Solar System would bring more results without needing to expose the explorers to severe risks and dangers, and at a fragment of the price. Why Mars? Why go the minimum of fifty-five million kilometres (average distance of 225 million kilometres) when it would have been easier to colonise Moon in the first instance? Of course, Martian proportions suggest this planet is a suitable target for human expansion. Although, after Earth, Mars is arguably the most hospitable planet in the Solar System, its habitability is an optical illusion and Mars is probably 'an awful place to live' (Cockell 2002).

The reason why the Red Planet becomes so appealing to contemporary Westerns, or at least to those who take an active interest in space exploration, is ultimately the romanticised vision of humankind to become a space-faring race. It is the pursuit of a dream of a few. Fantasies about interplanetary space travel and cosmological futures are being sold on our bookshelves, broadcast on our radios, screened at our cinemas, shared via our newsfeed. And somewhat unproblematically presented to us by the tycoons of space commerce as their aspiration inherited from generations of dreamers.

6.2 The Fictional and Scientific Histories of Martian Futures

Why has Mars captured the human imagination and became such a significant actor in the contemporary cultural repertoire? One cannot understand what is happening today unless an understanding the history has been achieved. The visions of exoplanetary futures, and in fact, all visions of space colonisation, are written in our science fiction stories. The answers are illustrated by Martian appearances in cultural texts and narratives that have approached the Red planet with different storylines, both with fascination and fear. In considering the text as ‘a site for creation of meaning’ (Berger 1997:13), our science and fiction books show that Mars holds a central post in the Western imagination. The new meanings are also created in press releases, where the promises, visions, dreams, grand ambitions and eloquent advocacy materialise into a phantasma of a Martian future.

Today a classical piece of the world’s literature, ‘The War of the Worlds’ by H. G. Wells was first published in (1898). This fictional template of the first contact has, in fact, a rich cultural history starting with a classic novel and continuing with many more screen adaptations. The film of that title from 1953 directed by Byron Haskins was released in the US and ten European countries between 1953 and 1955 (IMDb 2012a). Steven Spielberg’s movie was distributed to sixty-seven countries worldwide in 2005, showing on more than four-thousand screens in the US and UK alone (IMDb 2012b). Reaching a global audience, the fictional invasion of a superior alien race became a popular part of the modern narrative history of scientifically minded Westerners and transmitted around the globe. Martians became even more popular in 1939 when Orson Welles adapted ‘The War of the Worlds’ for a radio broadcast (Welles 1938). The latter on-air dramatisation of alien attack caused nationwide panic and demonstrated the ‘power of a narrative’ (Berger 1997:138) and the compelling power of mass media. These are powerful stories that bring the distant near and place the unfamiliar realities closer to the human condition.

The background of popular stories is to be sought in science, to be specific in the period astronomy. Italian astronomer Giovanni Schiaparelli described channels on Mars in 1877. The famous astronomer Percival Lowell, during his observations of the surface, described the canals and concluded that the Martian canals are of an artificial origin (Lowell 1903, 1909). Nikola Tesla (Musk’s alter ego) allegedly believed he detected a signal originating from Mars and that in fact, he had detected a message from the inhabitants of Mars. Mars had received great attention from former generations of scientists, being a logical first-hand close-by target along with the moon and Venus. Lowell’s speculations left enormous impressions on science fiction. Shortly after that, in 1913, Edmund Ferrier in his article ‘What Is Life on Mars Like?’ published in the *North American Review* summarised contemporary discoveries and discussed the evidence from Percival Lowell’s observations of Mars’s surface and Wells’ fictional story. This is, of course, a line of reasoning other scientists might have been reluctant to do. Ferrier, based on the knowledge to date, asked the following question:

> Does all this mean that there are no inhabitants in the planet Mars? No. Mars is certainly inhabited. The collapse of the fairy world constructed by bold imaginations on the base of the canals of Schiaparelli disposes only of the wonderful engineers of whom Mr H. G. Wells has given us, in his War of the Worlds, such a fantastic and captivating description. (Ferrier 1913:108).

In the early science fiction stories, the nearby celestial bodies were the first ones to be thought about as potential sites of exploration. 'Somnivm, seu Opus Posthumanum de Astronomia Lunari', known as 'The Dream', is scientific fantasy written by Johannes Kepler, published a few years after his death, in 1630. The book is considered to be the first work of science fiction (e.g. Christianson 1976). It comes as no surprise that the first planets considered as candidates for being inhabited were objects in our solar neighbourhood—that is the moon, Venus and Mars. Messeri explained how our attitudes about Mars are shaped by narrations created on Earth by science fiction authors, scientists, philosophers and politicians (Messeri 2016). Helmreich in his book 'Alien Ocean' described an analogy between the search for terrestrial life in extreme environments, such as extremophiles inhabiting deep-sea vents, and extraterrestrial environments that may potentially be a habitat for a life (Europa and ultimately Mars). It is worth noting that by doing so Helmreich also recorded the conceptual shift within the scientific search for life beyond earth, from intelligent to primitive life (Helmreich 2009). NASA's Europa Clipper spacecraft is being planned for launch in the 2020s to answer the question if a liquid water ocean beneath the surface of Jupiter's moon Europa could have the ingredients to support life.

Meanwhile in 2012, in a different part of the Solar System, the Curiosity Rover landed on Mars with the main scientific goal to help determine whether Mars could ever have supported life. The headlines, including the international journal of science 'Nature', described the landing operation as '7 min of terror' and the mission was discussed across numerous social media platforms, including Twitter and Facebook. Follows a first-hand description of the 'Mars Curiosity Rover Landing Broadcast at Times Square, Earth' as recorded and retold by one of the participants. The account captured the immediate reaction to the landing of NASA's Curiosity rover, during the online broadcast at Times Square, New York, was shared via his Flick profile:

> Tonight thousands gathered in Times Square to watch the Mars Curiosity rover safely landed on Mars after a 352-million mile, 8.5-month journey. Seeing so many people gathered on a late and rainy Sunday night was truly awesome to experience. People were captivated by the images being displayed on the Toshiba screen at Times Square and listened closely to the audio broadcast over their phones. When the rover landed, Times Square erupted in a loud cheer followed by chants of 'Science! Science! Science!' and 'NASA! NASA! NASA!' It was really incredible to see so many people collectively captivated by science and exploration. (Baraty 2012)

The account of the real-life experience is perhaps more than the collective expression of the 'explorer's complex' or shared enthusiasm for American space missions. However, another societal reaction has been recorded in a witty context. One online article, 'NASA's rover Curious sparks new wave of space memes', explains how humorous responses to the landing spread across the Internet and eventually made it to the TV news (such as Czech television). The photo-shopped image, which went viral on social networks, presents the Martians watching the Curiosity rover reaching the Red Planet. The little grey Martian natives are holding a large banner saying 'Yankee go home!' and 'Mars for Martians' (myjavier007's profile, Deviantart 2013).

Although no human had set his(her) feet on the planet, Martian commerce has already started to sprout and prosper. Moon Estates Ltd products include land on the Moon, Venus and Mars that is being sold online. An absence of space regulations that would guide both the governmental and private activities in space is one of the circumstances for the privatisation of celestial bodies. Possibly, then, NASA should reconsider its space mission budgets and landing plans and invest in an extraterrestrial real estate to secure a landing site, because the rest of the Martian surface will be privately held. Mars is popular; Mars is interesting. Mars is appealing even for the real estate's agencies. Mars is trendy. Mars is the next step.

6.3 The Martian Scientific Culture

The Martian culture, as a set of customs, integrated patterns of human knowledge, beliefs and social behaviour, will emerge in the new planetary settings, particularly in the case of the permanent or semi-permanent settlement. Speaking of culture, the term means different things to different people and is often confused with 'civilisation'. Culture is one of the two or three most complicated words in the English language (Williams 2014). The pages to follow argue that (a) the lives of the Martian astronauts, and (b) the lives of future colonists, will be defined and dominated by their science and technology, as the key tools of survival and a way of life.

In contemporary anthropology theory, term science as a culture or also 'cultures of science' refer to an approach of the social study of science. Sarah Franklin introduced the term 'cultures of science' (Franklin 1995:163) to capture the cultural contexts which shape and inform scientific practice. In one of the first ethnographies of scientific practice, 'Beamtimes and Lifetimes', a study of the world of high energy physicists, anthropologists Sharon Traweek identifies that one of the key characteristics that an anthropological study should offer is 'accounts of how the world is interpreted in cultures' (Traweek 1988). One of the conclusions of Traweek's study is the designation of what makes a culture. Here, Traweek described the world of high energy physicists, the culture of science, as an 'extreme culture of objectivity: culture of no culture' (Traweek 1988:162).

Recent astronaut selection and candidate choices explicitly state that science is the basic qualification requirement. To become a NASA astronaut today a bachelor's degree in engineering, biological science, physical science, computer science or mathematics is required. The International Space Station, a permanent outpost with six people on board, serves as a research laboratory with experiments being conducted in space. Similarly to the ISS, this will be the case of Martian station, and the survival of the Martian group will be dependent on the technology and imported supplies rather than the availability of local resources. If the process proceeds as planned, the spaceships will transport the settlers together with sophisticated science, advanced technology and first-class engineering across the vast distance and by doing initiate the new wave of space exploration.

Even when indulging in socio-science fiction and imagining what might be, the use of term 'culture of science' is relevant. The future Martians will resettle to a new planet and will need to interpret their lived world as well as the unexampled experience of the new world. Science and technology will be the key elements of survival, daily life and communication with the outside world. Traweek also pointed out to the importance of ecological settings and the developmental cycle, social organisation and cosmology of the group as four elements that make up the four domains of scientific community life. Traweek's study sets a strong theoretical and methodological background for any study that deals with the scientific community. One of the conclusions of Traweek's study is the designation of what makes a culture; the definition Traweek presented draws upon the work of Schneider and Geertz: 'the group's shared set of meanings, its implicit and explicit messages, encoded in social action, about how to interpret experience' (Traweek 1988:7).

There are reasons to believe, some type of such culture of science will be formed in extraterrestrial conditions. For example, Antarctica, Earth's southernmost and also the coldest, driest, and windiest continent, is only inhabited by scientists and the only human structures are the governmental science stations. When the British Antarctic explorer, Robert Falcon Scott set off the on the journey to the South Pole in 1910, he had brought a team with a scientific background. His Terra Nova Expedition had both the scientific and geographical objectives (Scott 2006). Good science and collection of data have been a central part of other expeditions with major areas of interest in meteorology, geology, astronomy and geography. American polar explorer and a pioneering aviator, Admiral Richard Byrd in an interview on the television journal Longines Chronoscope, described the role of Antartic regions for science. In his book 'Alone' the roles of making the meteorological observations and keeping the records were more pronounced as they played a crucial role in his own survival struggle (Byrd 2003, first published in 1937).

6.4 Scientific Culture and the Distorted Images of Science

In highlighting the role of science as a regime of life for future Martians, one must not forget the society that aims to send its representatives to a hostile, distant planet, and that 'awful place to live' (Cockell 2002). The society that has been formed and shaped by its science. As Harding described, the 'scientific culture' (Harding 1991) is the Western culture, where becoming and being 'scientifically literate' plays an important part. In many respects being able to read, comprehend, and even reproduce the idea of science renders Westerners inseparable from it. Moreover, this is a world in which the Westerners are presumed to be the scientifically literate 'knowers' who have a basic command of scientific concepts and conventions. The fact that we live in a scientific culture is illustrated by the fact, how crucial roles sciences play in our everyday lives.

Insights into popular attitudes towards space sciences and space exploration, in general, provide public opinion surveys. In the UK, Ipsos MORI popular opinion survey showed (Ipsos MORI 2014 UK) that 28% thinks that NASA Mars Rover will find evidence of life on Mars in 2014. This was also an increase in the 19% that thought it would happen during 2013. In 2018, the Pew Research Centre Survey found that 72% believed that it was essential for the United States to continue to be a world leader in space exploration, compared to 58% in a June 2011 poll (Cary and Strauss 2018). However, only 18% of respondents said that sending humans to Mars should be a top priority for NASA (Cary and Strauss 2018). In a 2013 survey sponsored by Boeing Company, 73% of Americans ages 18–24 are confident that humans will go to Mars by 2033, as reported in the Mars Generation National Opinion Poll, that measured the U.S. citizen support for the exploration of Mars (Holler 2013). Despite the public support for the national space programme, can we identify a consistent vision of a Martian future presented to us by science or do people base their thinking about Martian lives on a mix of science and fiction? Is there a clear distinction between science and fiction or rather they overlap?

Space, the final frontier. The place onto which are our dreams projected. The stage for our countless science fiction stories and of our misinterpretations. The historical and present fiction stories may have created almost an illusion of habitability and a vision of earthlike-ness. This is, of course, an optical illusion. A distorted image facilitated by visions of a few, but on a large-scale and possibly large impact. Contemporary media and writers seem to continue and extend on this unfortunate tradition. In 2016, the following description of Mars was published in an in-flight magazine Panorama:

> The Red Planet has mountains with icy caps on their tops, volcanoes, deserts, valleys, lake tracks, sand, rocks, and, as it turned out, rivers. (UAI Panorama, Ukrainian Airlines, March 2016)

This romantic description seems to ignore many facts about the actual properties of the planet. When reading the magazine, one may get an impression that Martian surface looks pretty much like Earth, except it does not. Similar common misconceptions can be seen in movies. Taking a shower. Unlikely. Waste of most precious source, required to sustain life and secure survival. Instead, rather the use of wet towelettes as practised by astronauts on the International Space Station. Two researchers walking through the corridor, discussing aspects of their work or just chit-chatting. Unlikely. Zero gravity field during space flight. Different perceptions of direction and several ways it affects the human body, including physiological changes (e.g. NASA's Twin Study) on the months' long way to Mars. Going for a walk on the planet? The surface gravity on Mars is only about 38% of the surface gravity on Earth. Rivers on Mars? Water in the ice caps perhaps on the poles, but rivers, as mentioned in the magazine, are rather the empty scratches in a shape of canals as we can see on the 1877 map of Mars by Giovanni Schiaparelli.

No matter how appealing the science fiction seem to the public they often convey incorrect notions or inspire false ideas about what challenges and complex issues the Mars habitation entails. As a matter of fact, Mars is nothing like Earth. Truly extreme environment that terrestrial analogue sites can only more or less accurately mimic. Fictions stories and narrations of the cosmic journey of humankind have such as strong bond in the society that they can be useful for science popularisation. Yet, especially the elusive concept of Martian habitability can become its greatest challenge.

6.5 Scientific Fictions and Tricky Visual Rhetorics

But also scientist can be tempted into projecting their hopes into scientific data. The vision of Mars and actual Mars are often two different things. One infamous example gives us the first photo of Mars released. Now known as colour controversies, the colour of Mars debate started in 1976 when NASA's Viking 1 became the first spacecraft to land on Mars. The colour controversies came down to a question of how our vision of Mars affects the colour or filter used. The first pictures sent back showed an Earthlike blue sky, leading some to hope that Mars could sustain life at the surface. While the unprocessed view would show an estimate of the natural colour that humans would see if they visited Mars, the processed version had interpreted the scene as if it were viewed under Earthlike lighting conditions. Carl Sagan, the Viking team member announced at a news conference that 'Despite the impression on these images, the sky is not blue. The sky is in fact pink, which is an OK colour,' he added (as reported in 'Color Controversies Started With Mars, Not With #TheDress' by the NBC News 2015). The image that shows the three versions of the same scene on Mars, captured by NASA's Curiosity rover, reflect three different choices that scientists can make in presenting the colours recorded by the camera, can be viewed at NASA/JPL-Caltech website.

The visual rhetoric of science used to convey and communicate the science concepts, such as the above-mentioned perceptions of the colour of the Martian atmosphere, can produce a false vision of the reality. The data visualisation technologies and the planetary imagination (Messeri 2016) can present fabricated visions of distant worlds and immersive futures. They can approximate the distant worlds enabling the audience to 'virtually witness' the spectacular and envisioned realities (Kirby 2003).

The peculiar example of distributing fabricated realities is, of course, supplies the Hollywood production. The latest science fiction films feature astronauts on Mars continuing the legacy of Curiosity and Opportunity rovers in search for evidence of ancient life, and so also the astrobiology concept embedded within the plot of the story. The Last Days on Mars (UK 2013) showed the fate of a group of astronaut explorers whom one by one perish while collecting specimens on Mars. For instance, Europa Report (USA 2013) depicts events that unfold after unmanned probes suggest that a hidden ocean and single-celled life exists on one of Jupiter's moons, and six astronauts embark on an ill-fated exploratory mission.

The latest science fiction survival film The Martian (2015) would be another example of how the future of space colonisation and astronautics are imagined. No matter how interesting the plotline, the science fiction Hollywood productions do not always aim to convey actual scientific theories or scenarios. Rather, they present fictional events and interpretation of facts at best. Some may even blame science fiction writers for deceiving the public by producing a false or misleading image of science and give preference to science documentaries that are based on scientific facts and available evidence. For example, Pierson noted that popular media 'can sometimes confuse bad science, or even pseudoscience, with the real thing' (Pierson 2006:481). From viewer's perspective, however, both tell a story. They are storytelling practices (Haraway 1988), the modern narratives if you like, that are embodied in other parts of popular culture: mass media, social platforms. They are used by space advocates and ambitious entrepreneurs alike.

6.6 Terrestrial History of Extreme Cultures in Extreme Environments

Make no mistake, despite heroic rhetorics and offerings of the next ultimate frontier, the colonisation of Mars will be a struggle for survival. The contemporary examples that can inform the planning of future missions to Mars are provided by the initially competing American and Soviet space programmes, and later the joint space station (ISS). Some useful lesson should be learnt from the Soviet space programme. We should not forget the names of Soviet cosmonauts, Georgi Dobrovolski, Viktor Patsayev, Vladislav Volkov who died while returning to Earth from the Salyut 1 space station, the only casualties who died in space in 1971. Since the beginning of manned space flight, 18 astronauts have died during the actual space missions (plus Apollo 1 crew). When considering the space race, it is to be noted that the Mars exploration is going to unfold in new socioeconomic climate, especially if privately funded.

More contemporary analogy when thinking about Martian culture is the International Space Station, the first orbital habitat, which is entirely dependent on the terrestrial supply of subsistence and material. The ISS has been in operation since 1998 and has a stable population of astronauts. The first-hand experiences with life and works on the ISS give us an indication of what the social microclimate during space flight involves, including the daily chores, rituals and cultures carried on the space station, even conflicts between astronauts on MIR. We can find more useful parallels in the studies of Martian habitat simulations (e.g. Bishop et al. 2010; Bubeev et al. 2014). There is a number of studies available, specifically focused on exoplanetary exploration, such as a recent mock Mars mission of the NASA-funded Hawaii Space Exploration Analog and Simulation (HI-SEAS), terrestrial analogue programmes, prototypes of Mars habitats, greenhouses, simulations of an artificial environment where humans can live, be sustained, and survive. Yet, the planetary habitat simulations were tested in terrestrial condition. Effects of other-planetary conditions, the impact of cosmic isolation on the human mind and body, when humans were removed from all environmental cues, were not examined in natural analogues, such as lunar surface, yet.

The historical events and accounts of maritime exploration, on the other hand, can provide useful guidelines—yet not manuals—on how the survival scenarios occur and unfold. There are several accounts of bold expeditions, related to finding sea routes or discovering. Some of the most reputed include the circumnavigation of Earth, search of the Northwest Passage, polar expeditions, Mount Everest. There are famed stories of the great age of exploration and life in extreme conditions. Such accounts also bring relevant perspective to the Martian debate. Notably, the science stations based in polar regions contribute to the study and understanding of both human welfare and the role of scientific thought. Antarctica as a continent is solely inhabited by scientists and the only human structures are the science stations. Although the terrestrial analogues of extreme environments and accounts of survival stories may give useful guidelines to abide, one must be attentive to details when drawing such an analogy.

Antarctica, the most unforgiving environment, the Earth's southernmost continent, resisted human exploration for centuries. Known to whalers who hunted in the southern waters, Antarctica as a continent became a centre of interest in late 1880. Since 1900 the region became a target for exploration, scientific enquiry, first bases were raised. Any life in Antarctica is exposed to light stress (light and dark cycles), low temperatures and freeze cycles amongst others. Similar to Mars, Antarctica has no indigenous human population. Its extreme weather conditions, pristine environment and limited resources make Antarctica a useful example of exploration in adverse conditions. Yet, it still has the magnetic field that protects from radiation, there is air to breathe, ice that can be melted into the water to drink, and limited access to natural sources of food.

Rear admiral Richard E. Byrd, the international hero who piloted the first flights over the North and South Poles, described his experience of Antarctic winter in his book 'Alone' (Byrd 2003, first published in 1937). In a small hut, isolated in the polar night for months of darkness with no hope of rescue until spring, Byrd began suffering inexplicable symptoms of physical illness caused by the faulty heater and serious carbon monoxide poisoning. The only connection with the outside world was a morse code transmission in the interval of three times a week. In his diary, nowadays a classic of polar adventure, Byrd describes the sounds of the scientific instrument as a reminder of the outside world. Another account shows that no catastrophic scenario training could prepare for a real-life scenario was Sir Ernest Shackleton's attempt to cross the Antarctic (1914–1917). Shackleton's ship Endurance was destroyed by ice and a small group set off on the epic journey to the South Georgia, to a remote whaling station to bring help to the rest of the crew who took refuge on the desolate Elephant Island. Shackleton succeeded and the story of unfortunate Imperial Trans-Antarctic Expedition is now known as an epic journey of endurance and survival (Shackleton 1920).

Antarctic regions as international space and environment serve as the model for international agreements on the exploration of the moon and other bodies in space.

In 1959, the twelve nations signed the *Antarctic Treaty*. The treaty was framed as an agreement so the continent 'shall continue forever to be used exclusively for peaceful purposes' and the second article 2 states that 'Freedom of scientific investigations and co-operation shall continue' (United Nations 1959). The treaty came into effect in 1961 and guarantees access and scientific research in all territory south of 60° latitude. Since then, the role of the Antarctic and outer space treaties in context with policies for scientific exploration and environmental protection had been studied (Race 2011) and Antarctica serves as a comparison to outer space as international spaces (Kerrest 2011).

6.7 The Typically Martian Problems of a Social Animal

To indulge in a social science fiction and envision the cultures beyond Earth, one needs first to disregard the arguments against the feasibility of Martian settlements as presented to date and assume that there is a reliable technology to keep the crew safe and sound through the journey to Mars, landing, settling down, establishing the colony; and—which is a crucial and mostly ignored question—bringing them back if needed. Leaving aside the medical and technological barriers, what are the first social and cultural difficulties we can sensibly predict the future Martians will encounter? What effect will live in a Martian bubble have on astronauts?

Anthropology offers various viewpoints on cultures. Traditionally, social anthropology focuses on the study of the predominant characteristic of a given group such as religion, cosmology, ritual activity and expressing customs. A relevant starting point is to ask what are aspects that do all human cultures have in common? Already in 1940, American anthropologist George Murdock published an article summarising the Cross-Cultural Survey, which gives a hint into key aspects of culture that are regarded universal (if one can speak of such thing given the variety of cultures on earth. Murdock concluded that 'culture is learnt, inculcated, social, ideational, gratifying, adaptive, and integrative' (Murdock 1940). For an exo-anthropologist, four key domains of social life will set the tone to the ethnography of the future colonies. As described by Traweek, those key domains of community life are ecology, social organisation, developmental cycle and cosmology (Traweek 1988).

The first truly interplanetary manned mission, however, will not be formed naturally but rather will be intentionally selected and carefully chosen by mission planners with regard to crew's physical, mental and intellectual capabilities and other skills. For example, Shackleton recruited for his Imperial Trans-Antarctic Expedition, which became one of the most known survival stories, 27-man crew of scientists, officers and seamen. There are many crucial social factors that will affect the group's culture, such as type of leadership (military command as in case of Astronautics), social organisation, group size, gender balance, developmental cycle, mobility, length of the stay and the ecological settings and type of subsistence (dependent or independent).

Indulging in social science fiction a bit further, one may speculate that more than the disruptions in natural cycles, sleep cycles, vitamin deficiency, monotonous diet, monotonous lifestyle, social claustrophobia and even cognitive distortion (Slobodian 2015) will challenge future Martians. It can be anticipated that some sort of an interplanetary desynchronosis, or space jet lag, may occur. The astronauts will be challenged to adapt to the new planet, the extraterrestrial experience. Another anticipated effect could be the Earth-sickness, the distress caused by being away from home. Being homesick, having preoccupying thoughts of home, feeling isolated and detached, and longing to return home, may especially be relevant for the permanent settlers. The Martian colonist will also be exposed to an overview effect of unknown force.

These are feelings and sentiments that belong to human nature. They are expressions of the human sense of belonging and sociability and play quintessential, although sometimes invisible or underestimated, roles in our lives. Aristotle in his Politics described a man as a 'naturally a political animal' (Book I, 1253a.2, Aristotle). The fact that humans are social animals is also evidenced by the universal need of a company or companions or the need for attachment (Rynearson 1978). Social ties, strong connections among humans that are used for sharing information, knowledge, feelings and experiences are the essence of being human. The social brain hypothesis tells us that human intelligence and cognitive responses evolved in response to the growth of social group (Dunbar 1998). The social cognition, a large number of social interaction seems to play some role in brain growth.

With limited social interactions, even no animals or earthlike environment to relate to, life on Mars is going to be lonely. The absence of the blue sky. Eerie silence (no birds, no trees, no sea). No animals or pets. Total reliance on the station, non-stop monitoring for safety reasons, and nowhere to escape. Extreme life in the extreme environment will be the one with the lack of social and environmental interaction, where human will be removed entirely from its natural world and live in an artificial bubble, no matter its sleek design and technological features. Those are indications of a large number of environmental, socio-psychological factors and valid reasons why the ethics of sending humans to Mars should be carefully considered.

The Martians are often envisioned as citizen scientists and it is anticipated that science be an essential part of a daily routine and way of life. If the mission to Mars will be of a scientific nature, such endeavour is likely to confer to the rules and values of scientific practice. This depends on the type of such an expedition that will unfold in terms of regimes of knowledge production (data collection) and likely dictate the daily routines of the Martian population. More details about the mission statement are required to speculate on the question. The composition of the team also will affect the group's cosmology. Concerning the survival and sustainability, one must consider also the economic viability of a colony. The crucial question is the economic viability of the station in case of private funding. Another factor is its dependence on external supply, as it is the case of the ISS. Likely, the Martian settlement will not be independent but will solely depend on supplies from the home planet.

It is also true that some of the Mars exploration plans are rushed with the critical details often ignored, and rather than feasible solutions they resemble utopia

(Slobodian 2015). Especially alarming are the visions of human settlement where the first generation of settlers the new generation of Martians since there no reliable data to assess the effect of exoplanetary environment on human foetus or newborn life). As distressing can be also described the use of enthusiasm as the main factor in overcoming adversity. Such as the following statement of Mars One on the journey to the red planet:

> Freeze dried and canned food is the only option. There will be constant noise from the ventilators, computers, and life support systems, and a regimented routine of three hours of daily exercise in order to maintain muscle mass. If the astronauts are hit by a solar storm, they must take refuge in the even smaller, sheltered area of the rocket which provides the best protection, for up to several days. The journey will be arduous, pressing each of them to the very limits of their training and personal capacity. However, the astronauts will endure because this will be the flight carrying them to their dream. (Mars One, Astronauts 2018b).

Although a good spirit and positive motivations are key in keeping the mental health of the team, underestimating extremes does not pay off. More than a dream will is required to survive. Resilience, skills, persistence, creativity, experience and a fair amount of good fortune. The unfortunate option that no one likes to admit is that the odds are that some will not endure. The lesson to be learnt from just the many survival stories, some of which were not accounted for simply because there was no one left to record them.

Speaking of the great era of exploration, dreaming about Mars displays historically rooted terrestrial cultural biases. In the popularisation process, Mars had become an icon of Western space aspirations and central idea of space settlement demagogy. A lunar base, for example, is a more logical step in the colonisation of the solar system and the establishment of a lunar colony could provide an essential experience with survival in the exo-terrestrial environment.

Mars had become yet another symbolic target of colonialism, imperialism and expansionism. Above that, the romanticised views of space travel, reconstruction of humans as multiplanetary species and other private space rhetorics do not solve the actual space flight problems, nor do they serve the higher purpose of the good of mankind. Nor does not the frontier pioneering or the 'moral law of the frontier' as rightly argued by Billings (2007).

6.8 The Envisioned Properties of a Martian Bubble

During the expansion of humanity into outer space, the future anthropology will have another opportunity to observe the emergence and the formation of culture in real time. To conclude, the broad sociocultural topics are as follows:

Imported culture(s) of science: To return to an earlier point and indulge in a socio-scientific fictional story about a Martian bubble, the prominent modus operandi for the Martian explorers will be the science. Since Mars is also the candidate planet of the search for life beyond our planet, the main goal of the journey will be looking for life. Basing the argument on the estimate that the team will be trained in science and engineering, skilled in survival techniques, and likely also diverse and international, one may suggest the emergence of a seemingly value-neutral scientified 'culture of no culture' (Traweek 1988:162). The courageous 'knowers' are the first step in the development of a Martian culture. If then we indulge in socio-science fiction further, one may anticipate the emergence of initially a high profile, scientific culture, driven by the ethos of discovery and survival. The length of stay, type of subsistence and crew's cultural dispositions, time management and even their autonomy will influence the social structure and relation to Mars as an environment.

Earth-boundness: Certainly, such culture of science and its human members will also rethink their relationship to Earth, and develop a new one. The overview effect experienced by astronauts is an overwhelming, intense feeling and a cognitive shift, in thinking about the self and the whole. Humankind is 'earthbound', and humans are evolutionary and culturally confined to earth. If we truly consider the effect the stay on Earth orbit has on astronauts, the inevitable question arises: how will the extreme detachment impact on emotional well-being and mental health of those who set their foot on Mars? Will they experience the very first Earth-sickness or interplanetary desynchronosis, an overview effect of an unknown gravity? That will be the ultimate out-of-Earth experience that no terrestrial analogue can fully mimic. Indeed, in becoming the multiplanetary species and during the new era of great explorers, humans will experience some sort of interplanetary space travel jet lag. They will be trained and expected to become extremophiles—organisms that thrive in extreme environments—themselves: Homo Extremophilaeus.

Sustainability of space research and future of space exploration: To date, the space activities, at least as far human space flight is concerned, have been almost exclusively an undertaking of governmental space programmes (American, Chinese, European, Russian). The situation, however, has changed and space commerce steadily rises on the importance and is becoming a major power in world space activities. Aside from the participation of private companies, again originating from already established space powers, another factor that should be appealed on is inclusiveness of space activities. White plans presented so far almost exclusively only include either science professional or public elite, some feel that we should focus on the inclusiveness and diversity of space exploration. Jasentuliyana, for example, already in 1993 mentioned that Mars exploration venture should involve all the countries of the world (Jasentuliyana 1993).

But it is not only the human lives we need to consider. It is the planet itself, as a symbol of how we relate to a treat our environment. One infamous analogy to the potential impact of commerce in outer space would be the commercial conquest of the ultimate terrestrial frontier, Mount Everest. Everest, in Tibetan as Chomolungma, was described by George Mallory as a symbol of 'man's desire to conquer the Universe'. The Earth's highest mountain is an alarming example of how far can the commercial high altitude tourism go: leaving a trail of garbage, oxygen tanks, waste and even human bodies behind. This is the cautionary tale to be taken seriously when thinking about going to Mars or creating any settlement on Mars or the moon, respectively. Although one may opine that on Mars, there will be no resources to waste and everything will be recycled or reused again. Still, the environment will be used, and any resource that can be acquired too. In context with the terrestrial environment and record of human history, here yet again comes the question if we speak about the exploration of Mars or exploitation of the planet. This hidden premise presents itself in the fact that Mars is already presented to us as a necessary backup option. We must leave our planet to survive, leaving a trail of garbage, oxygen tanks and perhaps even human bodies behind.

Martian social microclimate and steps to a Martian society: Rather than a giant leap, there will be small steps to a Martin society, even in a science operated mission. The future Martian anthropology will likely study how the science team was assembled, what were the aspirations and cultural background of the crew, how the team evolved during the flight and finally, how the explorers managed to and adapted to the extreme environment of the Red Planet, the journey back home or, eventually, their habitat. The stories about their life on Mars as ways of conveying meanings across space and time will be shared with other humans. Stories will be told about people who thrive in extreme conditions, who persist through challenges, stories about human cultural extremophiles. The success of the journey to Mars will depend on how the mission planners and the crew manage the technological challenges in environmental and social extremes.

In conclusion, before we respond to the cosmic longing for the ultimate adventure and send people to Mars, it is advisable to acknowledge and respect the human condition and make the physical health and mental well-being as a matter of utmost priority. Let's make sure the future Martians and their children will not become the guinea pigs, and make this the prime directive and principle any private space programme should adhere to. Before making the next proverbial giant leap, let's take every precaution and make sure the stories we read to our grandchildren one day will be those of a great adventure with a happy ending.

References

Aristotle. (1912). Politics: A treatise on government. Translated From The Greek Of Aristotle By William Ellis, A.M. London &. Toronto Published By J M Dent & Sons Ltd. & In New York By E. P. Dutton &. Co. http://www.gutenberg.org/files/6762/6762-h/6762-h.htm.

Arnould, J. (2014). The explorer's complex. *Space Policy, 30*(4), 185–189.

Baraty, Navid. (2012). Mars curiosity rover landing broadcast at times square, Earth. Flickr, Yahoo!, 6 Aug 2012. www.flickr.com/photos/vwmang/7724074220/in/photostream/.

Berger, A. A. (1997). *Narratives in popular culture, media, and everyday life*. Thousand Oaks: Sage Publications.

Billings, L. (2007). Overview: Ideology, advocacy, and spaceflight–evolution of a cultural narrative. In *Societal impact of spaceflight* (483–499).

Billings, L. (2017). Should humans colonize other planets? *No. Theology and Science, 15*(3), 321–332.

Bishop, S. L., Kobrick, R., Battler, M., & Binsted, K. (2010). FMARS 2007: Stress and coping in an arctic Mars simulation. *Acta Astronautica, 66*(9–10), 1353–1367.

Byrd, R. E. (2003). *Alone: The classic polar adventure* (p. 1937). First published in: Island Press.

Bubeev, Y. A. E., Gushin, V. I., Vasil'eva, G. Y. E., Vinokhodova, A. G. E., & Shved, D. M. (2014). Main findings of psychophysiological studies in the Mars 500 experiment. *Herald of the Russian Academy of Sciences, 84*(2), 106–114.

Capova, K. A. (2016). The new space age in the making: Emergence of exo-mining, exo-burials and exo-marketing. *International Journal of Astrobiology, 15*(4), 307–310.

Capova, K. A., Persson, E., Milligan, T., & Duner, D. (Eds.). (2018). *Astrobiology and society in Europe today.*, Springer Briefs in Astronomy Cham: Springer.

Cary, M., & Strauss, M. (2018). Majority of Americans believe space exploration remains essential. Pew Research Center: Internet, Science & Tech, Pew Research Center: Internet, Science & Tech, 26 July 2018, www.pewinternet.org/2018/06/06/majority-of-americans-believe-it-is-essential-that-the-u-s-remain-a-global-leader-in-space.

Christianson, G. E. (1976). Kepler's Somnium: Science fiction and the renaissance scientist. *Science Fiction Studies, 3*(1), 79–90.

Cockell, C. (2002). Mars is an awful place to live. *Interdisciplinary Science Reviews, 27*(1), 32–38.

Color Controversies Started With Mars, Not With #TheDress. Space Editorial, NBC-News.com, NBCUniversal News Group, February, 2015, www.nbcnews.com/science/space/color-controversies-started-mars-not-thedress-n314601.

Deviant Art. (2013).Yankee Go Home. www.deviantart.com/myjavier007/art/Yankee-Go-Home-408695673.

Dunbar, R. I. (1998). The social brain hypothesis. *Evolutionary Anthropology: Issues, News, and Reviews: Issues, News, and Reviews, 6*(5), 178–190.

Fairén, A. G., Parro, V., Schulze-Makuch, D., & Whyte, L. (2017). Searching for life on Mars before it is too late. *Astrobiology, 17*(10), 962–970.

Franklin, S. (1995). Science as culture, cultures of science. *Annual Review of Anthropology, 24,* 163–184.

Ferrier, E. (1913). What is life on Mars like? *The North American Review, 197*(686), 105–111.

Haraway, D. (1988). Situated knowledges: The science question in feminism and the privilege of partial perspective. *Feminist Studies, 14*(3), 575–599.

Harding, S. G. (1991). *Whose science? Whose knowledge?: Thinking from women's lives.* Ithaca: Cornell University Press.

Helmreich, S. (2009). *Alien ocean: Anthropological voyages in microbial seas.* Berkeley: University of California Press.

IMDb. (2012a). Release dates for The War of the Worlds (1953). http://www.imdb.com/title/tt0046534/releaseinfo.

IMDb. (2012b). Release dates for War of the Worlds (1985). http://www.imdb.com/title/tt0407304/releaseinfo.

Ipsos MORI. (2014). Britons give UKIP as much chance of winning the European elections as finding life on Mars. Retrieved from, www.ipsos-mori.com/researchpublications/researcharchive/3335/Britons-give-UKIP-as-much-chance-of-winning-the-European-elections-as-finding-life-on-Mars.aspx. Accessed 05 Dec 2015.

Jasentuliyana, N. (1993). Getting it right for Mars. *Space Policy, 9*(1), 5–9.

Kerrest, A. (2011). Outer space as international space: Lessons from Antarctica. *Science Diplomacy: Antarctica, Science, and the Governance of International Spaces* 133–142.

Kirby, D. A. (2003). Science consultants, fictional films, and scientific practice. *Social Studies of Science, 33*(2), 231–268.

Knapton, S. (2017). Human race is doomed if we do not colonise the Moon and Mars, Says Stephen Hawking. The Telegraph, Telegraph Media Group. Retrieved June 20, 2017, from www.telegraph.co.uk/science/2017/06/20/human-race-doomed-do-not-colonise-moon-mars-says-stephen-hawking/.

Koepsell, D. (2017). Mars One: Human subject concerns? *Astropolitics, 15*(1), 97–111. https://doi.org/10.1080/14777622.2017.1288512.

Kumar, S., Suman, S., Fornace, A. J., & Datta, K. (2018). Space radiation triggers persistent stress response, increases senescent signaling, and decreases cell migration in mouse intestine. *Pro-*

ceedings of the National Academy of Sciences, 2018, 201807522. https://doi.org/10.1073/pnas.1807522115.
Lowell, P. (1903). The cartouches of Mars. *Proceedings of the American Philosophical Society, 42*(174), 353–377.
Lowell, P. (1909). Mars as the Abode of Life. *Science, 30*(767), 338–340.
Mars One. (2018a). Roadmap—Mission. www.mars-one.com/mission/roadmap.
Mars One. (2018b). Mars One Astronauts—Mission. www.mars-one.com/mission/mars-one-astronauts.
Messeri, L. (2016). *Placing outer space: An earthly ethnography of other worlds*. Durham: Duke University Press Books.
Milligan, T. (2011). Property rights and the duty to extend human life. *Space Policy, 27*(4), 190–193.
Musk, E. (2017). Making humans a multi-planetary species. *New Space, 5*(2), 46–61.
Murdock, G. (1940). The cross-cultural survey. *American Sociological Review, 5*(3), 361–370.
Olson, V. A. (2012). Political ecology in the extreme: Asteroid activism and the making of an environmental solar system. *Anthropological Quarterly 85*, 1027–1044.
Persson, E. (2015). Citizens of Mars Ltd. In *Human governance beyond Earth* (121–137). Cham: Springer.
Pierson, T. (2006). SETI Institute as a model for managing interdisciplinary science. *Acta Astronautica, 58*(9), 478–484.
Race, M. (2011). Policies for scientific exploration and environmental protection: Comparison of the Antarctic and outer space treaties. In P. A. Berkman, M. A. Lang, D. W. H. Walton, & O. R. Young (Eds.), *Science diplomacy: Antarctica, science and the governance of international spaces* (pp. 143–152). Washington, DC, USA: Smithsonian Institution Scholarly Press.
Rynearson, E. (1978). Humans and pets and attachment. *British Journal of Psychiatry, 133*(6), 550–555. https://doi.org/10.1192/bjp.133.6.550.
Schroeder, J. E., & Tuttle, M. L., (1992). Investigation of possible causes for human-performance degradation during microgravity flight. Technical Report, Document ID 19920012102, NASA.
Scott, R. F. (2006). *Journals: Captain Scott's last expedition*. OUP Oxford.
Shackleton, E. H. (1920). *South: The story of Shackleton's last expedition* (1914–1917). Macmillan.
Slobodian, R. E. (2015). Selling space colonization and immortality: A psychosocial, anthropological critique of the rush to colonize Mars. *Acta Astronautica, 113,* 89–104.
SXSW, Source: Elon Musk: We Must Colonise Mars to Preserve Our Species after a Third World War—Video. The Guardian, Guardian News and Media, March 12, 2018, from, www.theguardian.com/technology/video/2018/mar/12/elon-musk-we-must-colonise-mars-to-preserve-our-species-after-a-third-world-war-video.
Traweek, S. (1988). *Beamtimes and lifetimes: The world of high energy physicists*. Cambridge: Mass. Harvard University Press.
United Nations. (1959). Disarmament treaties database: Antarctic treaty. https://disarmament.un.org/treaties/t/antarctic.
Valentine, D. (2012). Exit strategy: Profit, cosmology, and the future of humans in space. *Anthropological Quarterly, 85*(4), 1045–1067.
Wade Holler, M. Demographics have no impact on American's confidence that humans will go to Mars within two decades. Explore Mars, March 2013, www.exploremars.org/topic/boeing.
Williams, R. (2014). *Keywords: A vocabulary of culture and society*. Oxford University Press.
Williams, L. (2010). Irrational dreams of space colonization. *Peace Review, 22*(1), 4–8.
Wells, H. G. (1898). The War of the Worlds. Book one: The coming of the martians, chapter one: The eve of the war. Project Gutenberg: HTML e-book, 2004. http://www.gutenberg.org/files/36/36-h/36-h.htm.
Welles, O. (1938). The War of the Worlds. Radio Play. USA: The Mercury Theater on the Air. (Adaptation of the book by H.G. Wells). www.mercurytheatre.info. Accessed 30 Mar 2013.

Chapter 7
Colonising Mars. A Time Frame for Ethical Questioning

Jacques Arnould

Abstract As a result of NewSpace's initiatives and successes, Mars is now attracting a lot of interest in the astronautical community. The Mars One initiative is joined by Elon Musk's projects so much so that the first inhabited missions on Mars and its colonisation are at the centre of all interests, in a context of space resources utilisation. Besides technical and legal evaluations, an ethical and prospective evaluation of these projects is essential: is the conquest of Mars really "open" to the human species? Under what conditions could we develop and impose today? Because Mars was declared to belong to the common heritage of mankind, we are all responsible for its future.

7.1 NewSpace: A Fresh Impetus for Space

NewSpace. One word, not two. This is the term now generally used by space professionals and the media to designate what Jean-Yves Le Gall, President of CNES, the French space agency, describes as "the current revolution in the space sector, driven by innovation for applications" He adds that this revolution

> could have the unexpected effect of stepping up the pace of Mars exploration. Slated for 2040-2050 only six months ago, launch of the first crewed mission to Mars is seemingly getting closer by the day, with proponents now talking about 2030 or even 2025. Why? Because we are starting to see the benefits of the spectacular cost reductions in satellites and space launches made possible by a series of projects and increased launch and production rates. As a result, what was previously just a concept to send a spacecraft to Mars capable of supporting a four-to-six-person crew in space for two years is now on the verge of becoming reality.

The fact that these words were written in the editorial of a CNES magazine published in the summer of 2016 on the topic of Mars exploration does not suffice to explain Jean-Yves Le Gall's observation and analysis. The fact that one of the main European leaders in the space sector, reputed for his experience and cautiousness,

J. Arnould (✉)
The French Space Agency–CNES, Paris, France
e-mail: jacques.arnould@cnes.fr

K. Szocik (ed.), *The Human Factor in a Mission to Mars*,
Space and Society, https://doi.org/10.1007/978-3-030-02059-0_7

formulates this statement so clearly and publicly, before concluding that "[conquering Mars] is fast becoming THE new frontier for all of humankind[1]" tends to confirm that a new space venture is definitely afoot and that the red planet is being singled out.

Of course, NewSpace is first about facilitating access to space and initially to Earth's immediate environment. I am sure that NewSpace entrepreneurs, Elon Musk, Jeff Bezos et al., are familiar with this quote from Robert Heinlein, the author of the novella *The Man Who Sold the Moon*: "Once you get to low-earth orbit, you're halfway to anywhere in the solar system". Whether it is with a view to experiencing weightlessness and seeing the Earth from an altitude of 100 km during a suborbital flight (Virgin Galactic, Blue Origin), servicing the International Space Station (SpaceX) or launching a satellite constellation dedicated to Internet connections (OneWeb), the initial challenge is to access space at a cost that is compatible with the goals and constraints of private companies operating for profit. To achieve this, it is not enough to be free from the administrative and political constraints that bind public bodies. Rather, a new programme and team management methods need to be implemented and, above all, a sales and marketing policy must be developed that ensures that such flights around the Earth are profitable.

The second aim of NewSpace appears to be less immediate: to give fresh impetus to conquering space. Less immediate perhaps, but not absent: in September 2016 at the International Astronautical Congress in Guadalajara (Mexico), Elon Musk presented his own programme to reach and, between the lines, to create human settlements on Mars. I am in no position to assess the quality and technical feasibility of this Mars programme. I can simply note that Mars, which has been one of the main targets of scientific research in the space field for over 40 years, is now also the destination for ambitious settlement projects.

The promoters of these projects present themselves as true visionaries, explaining that they are not only doing this to (continue) to amass fortunes but also to fulfil a childhood dream to explore space. They are, moreover, attempting this with a somewhat different mindset from the one that prevails today in the "traditional" space sector. They claim to have reignited the pioneering spirit of the very earliest space ventures. Whatever ensues from accomplishing these dreams and projects, a new chapter of history is being written.

7.2 Strike Hard

Have times really changed? During the 1960s and 1970s, against the backdrop of the Cold War, means of communication were not of the same quality, scale or effectiveness as those of today. Yet, space was already the subject of well-considered communication; military operations were silenced; civilian activities relating to the sciences and human space flight, on the contrary, were the subject of an

[1]Le Gall, Jean-Yves. "Editorial", *CNES Mag*, July-August 2016, 5.

effective public awareness policy: remember the 500 million viewers who watched Armstrong's and Aldrin's first step on the moon in July 1969? Nowadays, careful consideration is still given to space-related communication: discretion prevails on the military side (apart from propaganda) and impacts on the civilian side; today's new communications resources can "strike hard".

The new entrepreneurs, a number of whom come from the internet sector, are building upon lessons from the past and present (e.g. successful planetary exploration missions by the space agencies) and modern technology. Whether it is with a view to proposing (or only envisaging) suborbital flights for tourists, who for the time being would need to be wealthy enough, holiday stays on board hotels in Earth orbit, flights to the moon to bring back pictures or leave objects there, or even trips to Mars, the shift from a national policy (for prestige or defence motives) to an entrepreneurial and commercial policy (to win over a market considered as sufficiently profitable) does not erase other dimensions. The scientific worth is systematically mentioned and even stressed; the desire to put on a show is clear (and pending actual accomplishment, this can even take centre stage) and lastly, perhaps above all, the existence of a genuine vision.

Let us take the example of Mars One. On its website, the company founded by Bas Lansdorp and Arno Wielders clearly state its goal from the outset: "Mars One aims to establish a permanent human settlement on Mars". Its schedule covers the start-up of crew training in 2018 and the first departure in 2032. All of this in the spirit of the missions of the 1960s since Mars One is presented as "the next giant leap for mankind", an explicit reference to Armstrong's declaration when he landed on the moon. Will the Mars One programme materialise? Or will it not? The challenges the Dutch company must face to accomplish its programme and, even more so, to keep to its schedule are clearly enormous. But they do not solely explain the interest and media hype surrounding its project. Admittedly, the one-way ticket proposed to successive teams of settlers may provide an explanation, albeit a morbid one. Yet this characteristic is self-evident: has there ever been a settlement (especially colonisation), in human memory, that has comprised repeated trips back and forth? Also, no one can deny the effectiveness of a programme carrying a vision that is both simple and daring, new yet already anchored in a deep human desire that blends exploration, expansion and exception. The weakness of this vision, however, apart from the selection of the crew members, is that it will probably be accomplished in a far more distant future than the schedule announced, which is particularly hypothetical.[2]...

The journey to Mars programme announced by Musk and presented in September 2016 is no less ambitious. Bringing it to fruition will also raise technological, financial and human issues. Musk does not lack vision either, not to mention the fact that Musk has acquired much greater competency in space transport than Lansdorp and his team. Those who were knocking at the doors of the great hall of the conference centre in Guadalajara seemed convinced of this and I imagine that many of them would have

[2]See https://www.sciencedirect.com/science/article/pii/S0094576515004294.

been ready and willing to board the immense spacecraft which the entrepreneur of South African origin presented with some impressive visuals.

Strike hard. Strike hard to impact our imaginations because colonising is no longer just exploring or conquering or exploiting. Colonising requires investment: human implication that is no longer temporary and can no longer settle for technological intermediaries. Humans will require convincing in order to leave behind their homeland, their Mother Earth, to settle in a world that is more unknown that any territory ever was for our ancestors, a world that is fundamentally and wholly extraterrestrial. Are we prepared to take such a risk?

7.3 Extreme Risk

On being asked about NewSpace, François Auque, formerly head of space operations at Airbus Defence and Space did not hesitate to explain, and rightly so, that "what is new is a pioneering spirit where failure is acceptable". This observation is all the more necessary considering that one of the most immediate and one of the most obvious temptations for space candidates is to minimise the difficulties that need to be overcome, the technological thresholds that need to be crossed and the boundaries that need to be pushed out. Hence, the book directed at encouraging "individuals like you to start thinking how you can build space colonies" (with explanatory diagrams no less) entitled *Nuts and Bolts of Space Colonization*. I wonder (since the book does not appear to provide any actual response) whether its author, Warren Reynolds Cozby Jr, is really serious or is this a work of science-fiction. Is he conducting a charm offensive, is he trying to convince himself or is this simply tongue-in-cheek? I neither possess the means to find out nor seek to judge the real intentions of Warren Cozby. I will just say here that the same questions could be asked about many speeches coming from the space field or relating thereto. And this holds true even going back many centuries.

Let us remind ourselves of what Johannes Kepler said. In 1610, having received from Galileo a copy of *Starry Messenger*, Kepler gave his response in his *Conversation with the Starry Messenger*:

> There will certainly be no lack of human pioneers when we have mastered the art of flight. Who would have guessed that navigation across the vast ocean is less dangerous and quieter than in the narrow, threatening gulfs of the Adriatic, or the Baltic, or the British straits? Let us create vessels and sails appropriate for the heavenly ether, and there will be plenty of people unafraid of the empty wastes. In the meantime, we shall prepare, for the brave sky-travellers, maps of the celestial bodies–I shall do it for the Moon and you, Galileo, for Jupiter.[3]

Intoxicated with the idea that mankind could one day escape its terrestrial prison, Kepler was convinced that now nothing was too high or too far for humanity to set its

[3]Quoted in Koestler, Arthur. *The Sleepwalkers: A History of Man's Changing Vision of the Universe.* London: Pelican, 1968. 378.

sights on reaching. History proved him right; however three and a half centuries later, the trip to space turned out to be more dangerous and less tranquil than imagined…

There is no point in listing here all remarks that were made thereafter in the same vein as Keplers' but let us give them credit for this at least, many of them were made with a conviction and sincerity similar to those of the German astronomer. Only times have changed, little by little, and the effective accomplishment of the first trips to space has forced us to be more realistic and more reasonable regarding projects and promises. Particularly those of NewSpace. While technological feats in the areas of graphics, visuals and staging now allow us to present upcoming projects and as-yet inexistent realities (to the point that we now have serious difficulties distinguishing between what is real and what is virtual or imaginary), it is necessary and even essential to remain vigilant as to the limitations, not to mention the dangers, of technology forcing. This expression designates the fact, the tendency or perhaps the temptation to expect technology to offer the solution to problems, dangers or disorders that it itself has raised. Technology forcing is a sort of outgrowth from the engineers' slogan "We can do it". A sort of "solutionism" (Luc Ferry) that is regularly mentioned and even put forward in relation to environmental crises or climate change. Technology forcing must not be systematically called into question or blamed. Human beings and the societies they make up, have managed to endure many crises during their history by virtue of their imagination and by developing technological solutions. But this singular ability must not become an alibi for ignoring or pretending to ignore the difficulties and the dangers, the errors and the inconsistencies of projects straight out of the imagination, even if the latter are presented as emergency exits—solutions to the catastrophes that threaten our humanity. Planet Mars and its colonisation are in no way a solution to the threats of global climate change. As former UN General Secretary explained in 22 September 2014: "There is no plan B because we do not have planet B", for all the humankind. Mars is an extraordinary place to pursue scientific discoveries about the planetary dynamics and about our own planet; but Mars will not be a planet B for us.

As for the risks relating to these activities, space is currently the stage for an encounter, or perhaps a clash, between two cultures, that embodied by Gene Kranz and that put forward by Elon Musk.

In 1995, when the Apollo XIII incident was made into a film by director Ron Howard, the scriptwriters came up with the idea of having Ed Harris, who played the role of Kranz, the flight director for this mission, say a phrase that quickly became cult: "Failure is not an option". In reality, when the Apollo XIII accident took place, and during the extraordinary rescue operation of the crew that ensued, Kranz never actually uttered those words; but he liked it so much and it captured the prevailing spirit that the NASA directors wanted as a team mindset so well that it became not only the title of Kranz's autobiography (published in 2000) but also a slogan that has been repeated and used over and over again on many NASA-released products. While coming from the mouth of Harris-Kranz, it sounds like a word of encouragement, an order, or even an incantation, it also conveys a reassuring message directed at political leaders and the public opinion in particular: everything, in NASA practices and culture, is undertaken and implemented so that failure never happens. This is a motto that other space agencies have also adopted.

We need to be clear: the possibility of failure is an intrinsic part of space ventures. If we were to identify the number of rockets that exploded upon launch, or compile a list of space probes that failed orbit insertion attempts around another planet or crashed there, or count the number of astronauts who tragically died on a space mission, this would only confirm what each of us already knows. And, the satisfaction with which those responsible for European launcher Ariane announced that the latter successfully completed its 82nd consecutive launch (December 2017) suggests that failure is still possible… yet, it must not be considered as an option. Let us make no mistake about it: such a position is not monolithic. It can be linked to risk philosophies that are very different from each other, depending on the era, company cultures or even engineer cultures. Or the difference between "Don't fly if it cannot be shown to be safe" and "Fly unless it can be shown not to be safe". Even so, the "official" line that failure is not an option prevailed up to the present in the space community.

Elon Musk has a very different culture from that of Gene Kranz. For the founder of SpaceX, failure is an option. He explains during an interview for a 2005 article in *Fast Company*: "Failure is an option here. If things are not failing, you are not innovating enough".[4] In other words, innovation is not an option. Regardless of the cost, I might add. And this applies to everything, to the point of being commonly accepted, or even a rule. When talking about plans for manned flights to and then on Mars, Musk does not hesitate to admit that going to Mars is dangerous: "It's dangerous and probably people will die—and they'll know that. And then they'll pave the way, and ultimately it will be very safe to go to Mars, and it will be very comfortable. But that will be many years in the future".[5] Has the culture of competitiveness replaced the culture of caution, like during the space race years? Or perhaps this indicates a pioneering culture.

Whatever it is, Musk's position is the one that prevails today in the American space entrepreneurship environment. Geneviève Fioraso notes this in her report on "The challenges and prospects of France's strategy in the space sector within Europe and in a global context" submitted to the French Prime Minister in July 2016. After observing the reactions to the failed launch of Musk's Falcon 9 in June 2015, she concluded: "In the United States, venture capitalists have even reversed logic and have turned the experience of failure into a positive criterion for selecting the project leaders they will finance".[6] Failure is considered part of learning. Nobody will say to those who fail: "We told you so!" Instead, they are somewhat encouraged: "Well done for trying!" A mixture of entrepreneurial spirit and extreme ambition that is sometimes even excessive: anyone who wants to think big and innovate must be prepared to face failure. Without forgetting to cultivate other qualities such as

[4]See https://arstechnica.com/science/2016/05/because-failure-is-an-option-spacex-can-do-stuff-like-land-rockets-on-a-boat/.

[5]See https://www.ibtimes.co.uk/elon-musk-people-will-die-first-spacex-missions-mars-1565387.

[6]See Fioraso, Geneviève. *Open Space. L'ouverture comme réponse aux défis de la filière spatiale* [Open Space. Openness as a response to the challenges of the space sector], 2016, 100.

courage, caution, patience and the capacity to weigh up information. These are essential qualities for colonising Mars and ethical questioning should also be added.

7.4 Why Human Settlement on Mars?

Exploration, conquest, human settlement: so many stances, so many movements that are not unique to the human race but characterise all living species, across all spatial and temporal scales. Do natural impulses suffice to justify a project or venture of the magnitude of creating human settlement Mars, without questioning its legitimacy? Clearly not. We are not only natural beings subject to many determinisms, but also cultural beings who have, or lay claim to, a certain freedom and, in this respect, we must think (to use Foucault‘s expression), in other words, examine the causes and effects of our actions, what has motivated them and their consequences. It is not certain that such questioning is still alive within the space community… An example I found is the report submitted by Norman Augustine to Barack Obama in October 2009.

Norman Augustine worked for the US administration at the Department of Defense and worked in the aeronautics industry and the space sector, where he successively occupied the positions of Chairman and CEO at Martin Marietta and Lockheed Martin. Did the US government take him at his word—“The optimum committee has no members”—when it asked him to direct the works of two committees responsible for examining the country’s space policy? *The Future of the United States Space Program* in 1990 and *Review of United States Human Space Flight Plans* in 2009: the two reports published by the “Augustine Committees” influenced and even reoriented US space policy, particularly the second one. On 1 February 2010, President Barack Obama indeed announced the cancellation of the Constellation programme initiated in 2004 by his predecessor, George W. Bush, a programme whose main aim was to send astronauts to the moon around 2020 for long-term missions. Whatever way this presidential decision was implemented (Constellation was not really cancelled…), three reasons were given to justify it: excessive overrun of the budget allocated to this programme, schedule overrun and, lastly, the absence of any real innovations associated with the project. It was a big blow for many US space industry players, but not for all. On the advice of the Augustine report, which believed that NASA should now rely more on support from private operators for low-Earth orbit operations, some of the funds released through ending the Constellation programme were devoted to this new direction in US space policy, in other words, NewSpace. On its level, the work of the second committee chaired by Augustine therefore contributed to the emergence, or at least to the development, of new space entrepreneurs, due to a partnership already in place with NASA. This report contains another particularity that is worthy of interest: it is perhaps the first official document, in the ordinary meaning of the term, that in its own way and style mentions and even promotes an ethical approach to be applied to space. Let me explain.

Several times, this text, in fact, proposes criteria that are simple yet effective for ethical appraisal and even decision-making in relation to space programmes:

> We explore to reach goals, not destinations. It is in the definition of our goals that decision-making for human spaceflight should begin. With goals established, questions about destinations, exploration strategies and transportation architectures can follow in a logical order. While there are certainly some aspects of the transportation system that are common to all exploration missions (e.g. crew access and heavy lift to low-Earth orbit), there is a danger of choosing destinations and architectures first. This runs the risk of getting stuck at a destination without a clear understanding of why it was chosen, which in turn can lead to uncertainty about when it is time to move on.[7]

It is not difficult to apply this method to the Constellation programme and to understand the US government's decision. Its destination was clear: the moon; its goals, on the other hand, were not sufficiently clear. Augustine and the members of his committee concluded that, in this case, it was better to suspend, cancel or at least profoundly rethink the Constellation program. Which is what the US President decided to do at the start of 2010, whatever the actual impact of his decision was on the continuation of national space operations.

Clearly, we cannot reduce all ethical considerations to the distinction proposed between the destination and the goal, which is indeed well suited to space ventures. But the related mindset seems to me to be relevant: rather than being distrustful and critical a priori, it is better to try identifying the deeper motivations behind a project or venture, its end purpose (whether goals and/or destinations) and assess its consequences, both expected and feared. This mindset is all the more relevant in today's world where, in the space community, in particular, the "We can do it" attitude of engineers readily prevails—the culture of performance that I mentioned above. However, a culture of utilitarianism also seems excessive and unsatisfactory. To stick with space technology, let us measure our societies' current level of dependency on satellite systems: we can no longer do without them for even a second, since their capacities to collect and send the data now required by most of our current-day activities have become so indispensable for us. Communications, meteorology, geolocation and observation are only the headliners of an ever-growing list of space applications. For all that, has the issue of the goal been resolved (that of the destination seems irrelevant and futile for these travellers who endlessly circle around in celestial orbits)? For the immediate future, no doubt; but it must be acknowledged that at a certain stage of technological development, it is no longer possible to know whether the purpose creates the tool or whether the tool creates the purpose. And then, at that stage, ethics amounts to being concerned about the social, political and cultural value of technology and practices whose existence can no longer be called into question, but only their goals.

This is indeed the ethical issue surrounding the colonisation of Mars: between pure performance (and there is still a lot of technological progress to be made) and the

---

[7] Review of U.S. Human Space Flight Plans Committee; Augustine; Austin; Chyba; Kennel; Bejmuk; Crawley; Lyles; Chiao; Greason; Ride. *"Seeking A Human Spaceflight Program Worthy of A Great Nation"*, http://www.nasa.gov/pdf/396093main_HSF_Cmte_FinalReport.pdf, 33.

prevailing utilitarianism (in what way would the red planet actually be useful for all humanity?), where is the best answer regarding the possible motivations behind such a project? Once again, is it enough to answer that this is a movement that is natural for the human species, like all living species, and because it is natural, it is a movement that should not be inhibited or prevented? It seems strange to oppose or impose the laws of nature to guide the choices of humans who have managed to become "like masters and possessors of nature" to quote René Descartes. Consequently, the question of human settlements on Mars (and perhaps colonisation) comes under the scope of reason. Not a theoretical reason but a practical reason rooted in a place (Earth) and time (ours).

7.4.1 *Gradatim Ferociter*

Gradatim Ferociter. Step by step courageously. The boots on which this virile and wise Latin saying is emblazoned do not resemble the famous moon boots of the Apollo missions. They come straight out of the American far west and could have graced the feet of a legendary cowboy, or one of his descendants on horseback. Jeff Bezos, who was given them by a friend, readily says that they bring him good luck and translates the Latin motto as "step by step, ferociously". Bezo's space career is well known.

After setting up Amazon "Earth's biggest bookstore" in 1995, the young American billionaire founded Blue Origin in 2000. As a schoolboy, he won a competition organised by NASA with an article entitled "The Effect of Zero Gravity on the Aging Rate of the Common Housefly" allowing him to visit the Marshall space centre in Alabama. Later, as a student, his valedictorian speech called to colonise space. Bezos explains that he is now fulfilling a childhood dream, inspired by the US space agency, and that, as a first step, he wants to offer and sell short trips to space at a price that is affordable for the greatest possible number of people. Six years went by before 13 November 2006, the first test of his spacecraft, New Shepard (a nod and homage to Alan Shepard, the first American to accomplish a suborbital flight on 5 May 1961). Thus, on 13 November 2006, north of the town of Van Horn, in a desert area of Texas, the so-called capsule only got 87 metres off the ground before coming back and gently landing on its launch pad. But this was the first step; the first of many steps towards colonising space. *Gradatim*, step by step.

Nine years later, on 23 November 2015, New Shepard, still unmanned, reached the frontiers of space, this time ascending to an altitude of 100.5 kilometres, at a speed of Mach 3.72. While the capsule, intended for future astronauts, came back down under the canopies of three parachutes, the rocket on the other hand used its engines to come back and land smoothly on Earth. The engineers even had it carry out a brief hover flight before placing it back down on the launch pad, which had also become a landing pad. Such a technological feat had never before been performed... apart from the red and white chequered rocket from Hergé's adventures of Tintin! On 22 January of the following year, the same rocket set off once again for the frontiers of space and became the first launcher of this type to actually have been reused; the operation was repeated on 2 April and 19 June of the same year.

Gradatim ferociter, step by step courageously. Bezos often repeats that it is possible to climb steps and reach subsequent stages sooner or later, but that it is impossible to ignore, avoid or skip any of them. It is therefore important to be as patient as courageous. It is perhaps also better to ignore that other possible translation of *ferociter*: "with arrogance". There is, without a doubt, a spirit of competition between engineers and space entrepreneurs in the field of reusable launchers but they are all only too aware of the challenges to overcome and the long road ahead to be inconsiderate or disrespectful towards their competitors.

Gradatim. Bezos does not lose sight of the goal he set himself: contribute to colonising space. And to achieve this, suborbital flight is insufficient. This was why, on 12 September 2016, he unveiled his next project: New Glenn, this time in reference to the first American astronaut to have flown around the Earth. A rocket standing 82 metres tall in its "short" version and 95 metres in its "long" version, the first stage of which will, of course, be reusable. The goal is clear: although none of his clients has flown yet and the profitability of the space tourism industry has yet to be demonstrated, Bezos does not lose sight of his objective—to go beyond Earth orbit. After gaining experience with New Shepard, he wants to play in the big league, where Elon Musk is already comfortably ahead of him. Bezos does not mince words: "In the long run, deliberate and methodical wins the day". Method and deliberation: another way to translate *gradatim ferociter*.

Bezos speaks about the long term. This expression merits our consideration, as it is rather rare in today's business environment… and even in today's space sector. Let us recall the challenge launched by John F. Kennedy on 25 May 1961, which was barely three weeks after Alan Shepard's flight: he expressed the intention of committing the nation to achieve the goal of landing a man on the Moon and returning him safely to Earth before the decade was out. In other words, an extremely short time period in light of the technological capacities of the time. This challenge was taken up and achieved by NASA within the (short) deadline imposed, under the Gemini and Apollo programs. It remains a historical and cultural reference, even half a century later. Many people noted, regretted and criticised the fact that Neil Armstrong's step was not followed up by a real giant leap, a decisive breakthrough for humanity, and that human exploration of space did not go beyond lunar orbit due to a lack of momentum, lack of enthusiasm and perhaps also an absence of long-term vision. It is true that the moon was an extraordinary destination but unfortunately, it was also an end in itself, and voyaging there and back did not sufficiently or necessarily open up the possibility of taking another step further.

Perhaps we should consider projects or programmes that could be carried out in stages, step by step, each of them having its own goal and consistency, without it being imperative or essential to refer to a longer term. For example, those who develop programmes for human missions on Mars readily defend initial exploration of the red planet's natural satellites, Phobos and Deimos: goals that could already suffice in themselves but also coherent steps prior to the arrival of men on Mars. Can the same be said for technology development? Mastering new technology (I am thinking of nuclear propulsion) can be an iterative process: whatever the final result, the completion of each step can result in knowledge and expertise being acquired, and perhaps even breakthrough technology.

Gradatim ferociter: I think that the inscription on Bezos' boots could very well be wise advice for future space projects.

7.5 Colonisation of Mars: A Horizon for Humanity

I am convinced that Mars, its exploration and why not its colonisation constitute a horizon for our humanity and apprehension about its future. A horizon—perhaps an achievable goal—but certainly not simply a destination, as the moon was at the end of the 1960s. We need to leave behind once and for all purely technological projects that are simply missions similar to that of Apollo XI: to arrive safe and sound on Mars, raise a flag there and come back as quickly as possible only never to return; this would be a waste in financial, human and scientific terms. It would be a real ethical failure.

On the contrary, envisaging the colonisation of Mars can be an opportunity to start assessing our capacities and weaknesses right now, along with our knowledge and ignorance, as well as our limits, whether those imposed on us or those we impose on ourselves. This is what I call a horizon: a moving boundary that remains unreachable and inaccessible until we accept to take a side step and change dimension. This is what astronauts experience when they move away momentarily from the Earth's surface and pass above Earth's horizon. The colonisation of Mars will require a much greater leap to enter into a new perspective, a more extraordinary dimension. Earth images taken by robots on Mars only give a vague idea of this step to be taken.

Yet, even while remaining "under" this horizon, that is, within the framework of projects, we can already see the questions that we will have to address. I am choosing to mention three of them.

The first issue is the status to be given to the red planet. For the moment, space law declares that celestial bodies belong to the common heritage of humanity; a status that is called into question by national initiatives relating to projects to exploit the mineral resources of space. However, is what could be envisaged, or is currently being envisaged, in the ultimately limited overall framework of these projects suitable for human colonisation? Probably not. Consequently, should we not presently consider the horizon of colonisation in order to attempt to find a legal solution to the issue of exploiting space?

The second issue concerns human beings more directly. I already mentioned the challenge of apprehending danger and risk with the pioneering spirit. But will these not be different according to the methods chosen to allow humans to live on Mars? Building totally closed bases on Mars (like Biosphere II), undertaking to transform conditions on Mars through terraforming, envisaging to modify humans at the same time as their Martian environments (to make them "Martian transhumans"): how far are we prepared to go in transforming Mars and transforming human beings to live on the red planet?

The third issue has as its horizon the human species as a whole. We have never had such a clear sense of this, thanks to the development of all sorts of communications. Thanks to them or because of them, there is not much that we do not know about each other anymore. Who will be the Mars settlers? An elite who will be given the mission of saving the human species condemned to disappear from Earth (due to its own fault) or, on the contrary, and as has often been the case in the history of humanity, people who are marginalised for various reasons?

I do not have the answer to any of these three questions and this is why I spoke of a horizon: several possible answers can be given for them, since they are open to discussion and debate. Even so, they are not in vain, quite the contrary. Each of them, in its own way, offers an opportunity for humans to simultaneously prepare for their own future, that of the Earth and that of space, which they are and will be capable of reaching. Envisaging the colonisation of Mars is simply the act of continuing to explore the human condition.

References

Arnould, J. (2011). *Icarus' second chance. The basis and perspectives of space ethics*. Wien/New York, Springer (ESPI: Studies in Space Policy).

Arnould, J. (2017). *Impossible Horizon. The essence of space exploration*. Adelaide: ATF Press.

Milligan, T. (2015). *Nobody owns the moon. The ethics of space exploitation*. Jefferson (North Carolina): Mc Farland & Cie.

Pasco, X. (2017). *Le nouvel âge spatial. De la Guerre froide aux New Space*. Paris: CNRS Editions.

James, S., & Tony, M. (Eds.). (2016). *The ethics of space exploration*. New York: Springer (Space and Society).

Vance, A. (2015). *Elon Musk: Tesla, SpaceX, and the quest for a fantastic future*. New York: HarperCollins Publishers.

Chapter 8
Can Deep Altruism Sustain Space Settlement?

Jacob Haqq-Misra

Abstract Space settlement represents a long-term human effort that requires unprecedented coordination across successive generations. In this chapter, I develop a comparative hierarchy for the value of long-term projects based upon their benefits to culture, their development of infrastructure, and their contributions to lasting information. I next draw upon the concept of the time capsule as an analogy, which enables a comparison of historical examples of projects across generational, intergenerational, and deep time. The concept of deep altruism can then be defined as the selfless pursuit of informational value for the well-being of others in the distant future. The first steps toward supporting an effort like space settlement through deep altruism would establish governance and funding models that begin to support ambitions with intergenerational succession.

8.1 Introduction

The prospect of human settlements on Mars is becoming increasingly technologically feasible. Efforts by SpaceX, Deep Space Industries, Planetary Resources, and other private space corporations now fall in rank with government space agencies such as NASA, ESA (European Space Agency), JAXA (Japan Aerospace Exploration Agency), IRSO (Indian Space Research Organization), RFSA (Russian Federal Space Agengy, or Roscomos), and CNSA (China National Space Administration). Many of these corporate and government entities are developing successive plans to visit asteroids or Mars within the next few decades (Seedhouse 2010; NASA 2015), which are beginning to show prospects for economic gain in addition to scientific return. These recent developments all suggest that Elon Musk's vision of human civilization becoming a "multiplanetary species" (Musk 2017) could be realized in the coming centuries.

Technological advances that enable humans to settle on another planet or extract resources from planetary bodies must be matched by parallel advances in civilizational

J. Haqq-Misra (✉)
Blue Marble Space Institute of Science, 1001 4th Ave, Suite 3201,
Seattle, WA 98154, USA
e-mail: jacob@bmsis.org

K. Szocik (ed.), *The Human Factor in a Mission to Mars*,
Space and Society, https://doi.org/10.1007/978-3-030-02059-0_8

ethics. A lack of moral progress risks the danger of perpetuating the problem of the commons and other harmful colonial attitudes as human civilization ventures into space. Land use policies on Mars must account for the finite extent of resources while also respecting the non-appropriation principle of the Outer Space Treaty that restricts national claims to sovereignty (Haqq-Misra 2015). New governance models, either by modification of the Outer Space Treaty or by the creation of new international institutions, can promote solutions to the equitable sharing of Mars and other space-based resources (Ehrenfreund et al. 2013; Cockell 2015; Haqq-Misra 2016; Bruhns and Haqq-Misra 2016); however, implementation of any such idea would require commitment and cooperation from most of the major space-faring nations as a minimum. Nevertheless, establishing sustainable space settlements will require that humanity begin advancing its ethics in tandem with technology, prior to the arrival of the first humans on Mars.

Innovative approaches to fundraising, charitable giving, and other means of financing can further enable long-term and intergenerational initiatives such as space settlement. Corporations like IBM, Lloyd's of London, and the Swedish National Bank all hold experience in maintaining business tradition and financing ventures over decadal timescales and longer, while organizations like the Rockefeller Foundation and the Carnegie Corporation have promoted the advancement of knowledge through charitable contributions for more than a century. Crowdsourced models could also provide sustained funding for space settlement or other long-term scientific initiatives, such as the search for extraterrestrial intelligence (SETI). Specialized financial products could be tailored toward individual scientific objectives or development goals, such as a lottery bond debt security, which could provide a regular stream of income to ambitious projects while also providing consumers with a direct return on investment (Haqq-Misra 2018). The success of commercially driven space settlement will inevitably require tremendous financial foresight that could benefit from a donor, group, or crowd willing to invest in the distant future of humanity.

In this chapter, I define the concept of "deep altruism" as ambitious human efforts with high informational value across millennial timescales that do not provide direct benefits to the initial benefactor. I begin by drawing upon the analogy of a time capsule, which represents an abstraction of a long-term human effort with the intention of providing value to future generations. I then define a relative value scale based upon cultural, structural, and informational motivations. I next consider historical and contemporary examples of the completion time for projects across generational, intergenerational, and deep time. I conclude by discussing the feasibility of establishing a space settlement based upon a deep altruistic funding model.

8.2 Time Capsules

The time capsule serves as an example of a long-term effort with altruistic intentions, which highlights some of the unique challenges in approaching altruism across deep time. The Oxford English Dictionary defines "time capsule" as "a container used to store for posterity a selection of objects thought to be representative of a particular moment in time." Time capsules represent an attempt at preserving value from a particular time in history for the benefit of other people in the future.

The concept of a time capsule today usually refers to an object constructed with intention, in order to purposefully commemorate a particular time by preserving the value of its memory. Archaeological discoveries also provide scholars today with new information, but such discoveries rarely encounter concerted efforts of cultures from the past attempting to communicate with us today. The first deliberately constructed time capsule, the "Century Safe," was featured at the 1876 World's Fair in Philadelphia, which established the modern tradition of intentional time capsules.

Time capsules can further be distinguished by the intended future audience, which corresponds to the length of time that the capsule must remain preserved. "Target-dated" time capsules specify a particular length of time (e.g., a century) when the capsule will eventually be opened; "deliberately infinite" time capsules are preserved in perpetuity until future generations eventually decide to open the capsule (Jarvis 2002). Both types of capsules can suffer from the lack of complete information, such as including toys or trinkets without explanation of their significance, which limits the value of the objects to historians. An ideal time capsule constructed today, to maximize future value, should include a "full set of cultural-technical information drawn from the whole of human world culture" with a "10,000-year target span date" (Jarvis 2002). Other features for an effective time capsule include the existence of redundant copies to protect against loss, as well as electronic access to the contents in order to increase transparency and maintain interest.

A time capsule with a target date that approaches millennial scales or longer serves as an illustration of a general altruistic effort that operates over deep time. Short-term time capsules with a target date on the scale of decades could provide direct benefit to people who were living when the capsule was sealed. A longer term time capsule with a target date on the scale of a century would not necessarily benefit the individuals who sealed the capsule, but their grandchildren or great-grandchildren would likely benefit from the value of opening the capsule. A time capsule with a target date on a deep time scale of millennia or longer would represent a genuine altruistic effort, as the individuals or community that sealed the capsule would likely have no direct connection to the people opening the capsule. A time capsule motivated by a sense of deep altruism would seek to provide relevant and contextual information to humanity's distant descendants.

The problem of preserving the contents and knowledge of a time capsule over millennia poses similar challenges to other long-term projects attempted across history. The settlement of space, including the development of permanent human habitats and even the terraforming of an entire planet, will require unprecedented cooperation and coordination across deep time. As with the time capsule, any long-term effort like space settlement must maintain informational relevance and sustain operations across generations in order for its future value to be realized.

8.3 Cultural, Structural, and Informational Value

In order to further unpack the concept of deep altruism, I define a framework for comparing the relative benefits derived from long-term human projects. The investment cost in a long-term project is not necessarily a reliable indicator of its realized value, as seemingly inexpensive items may show themselves to be priceless when retrieved by its future recipients. Apparent success of a project today is also not a reliable indicator of future value, as human civilization will likely evolve its infrastructure and preferences across the next millennium. Instead, I draw upon value theory to establish a comparative scale for long-term projects based upon the ultimate type of value realized by future generations. Value theory is a broad approach in ethics for defining relative degrees of goodness, benefits, or other desirable features, which enables comparison of the relative value of objects or actions even if further quantification is difficult. For this analysis, I develop a specific hierarchical approach toward valuation of long-term projects based upon cultural, structural, and informational value. This value hierarchy is summarized as a pyramid chart in Fig. 8.1, with culture at the base, structure in the middle, and information at the apex.

The first level of cultural value refers to a long-term project that is primarily concerned with aesthetic factors, the preservation of tradition, and other features central to the group's identity. A project motivated by cultural value would seek to enable future descendants to experience similar, or greater, appreciation of the originating culture. Cultural value resides at the foundation of the value hierarchy because culture is inherent to all human activities. Thus, the remaining levels in the value hierarchy necessarily include culture in addition to other value considerations.

Structural value is the second level, which describes a long-term project that is primarily concerned with the preservation of materials, buildings, or other engineering feats. Structurally motivated projects require diligent maintenance in order

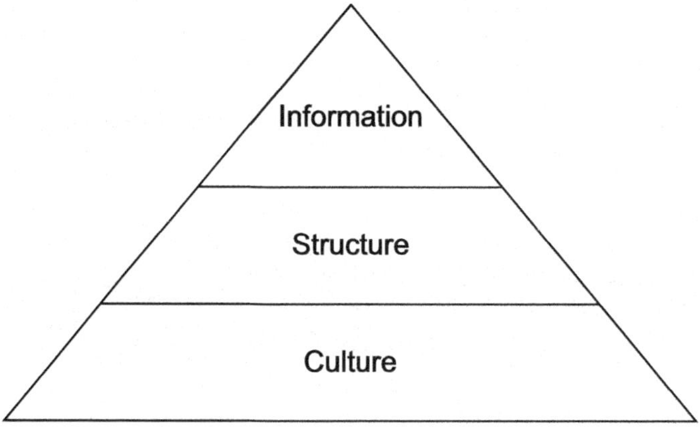

Fig. 8.1 The hierarchical relationship between cultural, structural, and informational value provides a framework for comparing the relative value of long-term human projects

to preserve the integrity of the original construction for future descendants. Many structures are intended primarily for utilitarian purposes, such as shelter or storage; the long-term preservation of such structures can be accomplished by active efforts by governments, individuals, or private organizations. However, some structures are more strongly linked with culture, such as religious temples, burial sites, or ancestral shrines. The value hierarchy therefore includes an intermediate level on the culture/structure boundary to account for structures that are maintained by time-tested cultural traditions.

The final level of informational value represents the realization of culture and structure to provide long-lasting benefits to human knowledge. A project motivated by informational value would seek to build upon knowledge traditions to enable future generations to solve major problems. Informational value provides tools and methods that enable solutions, whereas cultural value only provides a means of preserving information. Some projects involve significant engineering and management innovation in order to support the acquisition of knowledge, as is increasingly required by "big science" projects in physics, astronomy, and materials science. The value hierarchy therefore includes an intermediate level at the structural/informational boundary to account for physical structures that support efforts at achieving new knowledge.

This three-tiered hierarchy of cultural, structural, and informational value provides a relative scale for ranking long-term human projects across history. The purpose of such a relative value scale is not to judge the historical merits of any of these efforts; instead, this approach enables an analysis of the factors that enable the successful preservation of each type of value across deep time.

8.4 Completion Time

History is abundant with examples of long-term projects conducted across generational, intergenerational, and deep time scales. The completion time for such projects is defined as the amount of time between the initial conception of the idea and its final execution, analogous to the duration of a time capsule. Some projects include a target date for completion after a finite amount of writing, construction, or analysis—similar to a target-dated time capsule. Other projects are deliberately maintained in perpetuity, with no intention to cease operations, comparable to a deliberately infinite time capsule.

A range of historical and contemporary examples of long-term projects is shown in Fig. 8.2, with completion time on the horizontal axis and value on the vertical axis. Figure 8.2 is intended to be an illustrative, rather than comprehensive, collection of projects conducted over generational, intergenerational, and deep time with cultural, structural, and informational value motivations. This visualization of successful long-term projects enables a comparison of the factors required for sustaining altruistic activity over deep time.

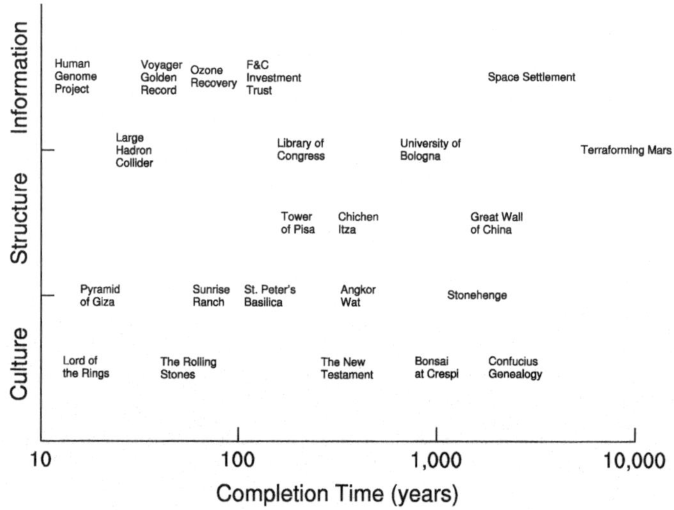

Fig. 8.2 Examples of long-term human projects plotted as a function of value versus completion time, where value is categorized as cultural, structural, or informational

8.4.1 Generational Time

Projects that occur within a single generation have a completion time between about 10 to 100 years. Generational time efforts often have a target date for completion, usually with the intention of finishing within the lifetime of the project's originator. The *Lord of the Rings* trilogy by J. R. R. Tolkien was written in stages over a period of 12 years, which represents a decadal-scale work of cultural value that still persists in completed form today. The Great Pyramid of Giza is an example of a work of structural value, which took about 20 years to build. Informational projects can also have a finite duration, such as the 30-year construction time to build the Large Hadron Collider, the world's most powerful particle accelerator. Likewise, the Human Genome Project required 13 years of collaborative research to obtain a complete mapping of all human genes. Successful generational time projects with a finite target date usually allow the originator to personally experience the full value upon completion, even if the final product persists for much longer.

Some generational projects are deliberately infinite by design and could extend into longer intergenerational timescales. The Rolling Stones is the longest continuously performing rock band, founded in 1962 and still performing with some of its original members; conceivably, the franchise could persist beyond the lifetime of the individual band members. One of the longest-running intentional communities in the US is Sunrise Ranch, operated by the Emissaries of Divine Light since 1945. The golden records on-board both Voyager spacecraft, launched in 1977, contain encoded knowledge about Earth and its inhabitants that will continue drifting through interstellar space indefinitely. The ozone hole is slowly recovering as the human use of chlorofluorocarbons (CFCs) has declined, but projections indicate 70 years or longer before the Antarctic ozone layer recovers to pre-1980 levels. Deliberately infinite projects necessarily begin at the generational scale, but if they are successful then they will continue beyond the lifetime of the originator to be tended by the next generation.

8.4.2 *Intergenerational Time*

The completion time for intergenerational projects is about 100 to 1,000 years. Intergenerational time efforts may have a goal or objective but not necessarily a reliable target date for completion. The time between the writing and canonization of the New Testament was about 350 years; although the text has been faithfully preserved, the original authors of most of the New Testament remain unknown. Many cathedrals, such as St. Peter's Bassilica in the Vatican, took a century or longer to construct, while the Angkor Wat temple complex in Cambodia was built in stages over a period of 400 years. The Tower of Pisa took 199 years to build in three stages, showing evidence of sinking with the completion of each floor. The Chichen Itza building complex of the Maya took about 400 years to construct, which remains a draw for tourists today. Many intergenerational structures and cultural artifacts still provide value for people today, even if for different purposes than intended.

Other intergenerational projects have maintained deliberately infinite operations, with no defined point of completion. The Library of Congress is one example of a robust infrastructure that has sustained the acquisition and cataloging of human knowledge for over 200 years. The 150-year-old Foreign and Colonial Investment Trust holds the record for being the longest-running investment fund. Educational institutions such as the University of Bologna and Oxford University have managed to sustain their structures and operations for nearly a millennium. Institutions that maintain deliberately infinite operations across multiple generations, while successfully adapting to change, will eventually approach the threshold of deep time.

8.4.3 *Deep Time*

A project that operates within deep time has a completion time of 1,000 years or longer. Deep time efforts must contend with dramatic shifts in geopolitics, changes in the Earth system, and other factors that remain less volatile at shorter timescales. Structures such as Stonehenge were constructed in stages over about 1,500 years, perhaps with a general goal but likely no target completion date. The Great Wall of China was built, extended, and repaired by multiple dynasties over about 2,000 years, with the goal of protecting the northern border from invading armies. Finite duration projects that operate over deep time represent long-lasting organizations that were able to maintain culture and engineering tradition over the course of drastic historical changes.

Deliberately infinite deep time projects require careful attention by successive generations in order to preserve knowledge that would otherwise be forgotten. Unlike structural value, which can persist even if the founding culture becomes extinct, cultural value can be lost if not preserved through written and oral tradition. The Ficus Bonsai Tree at Crespi, Italy, is among the oldest in the world and has received daily care for the past 1,000 years. Similarly, the genealogy of Confucius has been dutifully preserved for 2,500 years, enduring through numerous dynasties and political revolutions. Such deliberately infinite efforts at preserving culture seem likely to continue for the foreseeable future, as long as people continue to recognize their value.

Few, if any, human ambitions have successfully managed to realize informational value across deep time. Contemporary efforts to enable the human settlement of space represent a deep time ambition to achieve informational value. Space settlement is an infinite duration project that will likely be developed in stages, with an idealistic goal of enabling self-sustaining human populations that no longer require support from Earth. The timescale for achieving an initial human presence in space may be generational, but any long-lasting space settlement must successfully traverse deep time in order to demonstrate its autonomy. Even more audacious ideas to terraform a planet like Mars, so that it could sustain plant life and a breathable atmosphere, represent a deep time effort with a finite target date; however, transforming an entire planetary system would take 10,000 years or longer of patient monitoring in order to reach the desired climate state. If any such plans for the permanent human settlement of space actually do begin to take shape, then they will represent the first intentional effort at pursuing informational value across deep time.

8.5 Deep Altruism

I can now revisit the definition of deep altruism, drawing upon the relative value scale for long-term projects and the discussion of completion time. The word "altruism" is defined by the Oxford English Dictionary as "disinterested or selfless concern for the well-being of others, especially as a principle of action." In the context of long-term projects over deep time, altruism refers to selfless concern for the well-being of others in the distant future. The preservation of cultural tradition can include altruistic elements, but such efforts do not necessarily foster new methods of systematically increasing total well-being. Conversely, altruistic pursuit of informational value can expand knowledge and enable solutions to problems that significantly improve the quality of life. Deep altruism can therefore be defined as shown in the box below.

Deep altruism is the selfless pursuit of informational value for the well-being of others in the distant future.

The assortment of projects in Fig. 8.2 show a lack of examples with informational value and millennial completion time, although such an approach is necessary for the daunting task of settling space. However, many projects beginning today are motivated by altruistic intentions, with target dates that approach deep timescales. The Clock of the Long Now is being constructed with the intention of keeping time for 10,000 years; the clock is being funded by Bezos Expeditions and currently resides on land owned by Jeff Bezos. The Letters of Utrecht is a collective poem carved into cobblestones along the city's streets, with a new letter added every Saturday. The Letters of Utrecht began in 2012 and is intended to continue in perpetuity—or as long as the citizens of Utrecht permit. A record-setting organ performance of "As Slow as Possible" composed by John Cage is underway at the St. Burchardi church in Germany, with the first note beginning in 2001 and the piece reaching a finale in 2640. Although all of these efforts presently have been able to secure enough resources to maintain operations, they are still in the initial generational phase where the project originators are still alive and involved. The long-term success of these and other altruistic efforts will require effective succession between generations in order to transition into intergenerational and deep time.

Why would an individual or organization choose to engage in deep altruism? Reciprocal altruism features in many instances of biology, as natural selection pressures can operate against individuals who choose selfish or cheater behavior in cooperative groups (Trivers 1971). Non-reciprocal altruism seems to be a unique feature of humans (and possibly a few other primate species), with a less obvious explanation for the evolutionary benefits of selfless concern for even complete strangers. Understanding non-reciprocal altruism remains an ongoing area of research, with some analyses suggesting that social distance can correlate to expectations of reciprocal altruism (Brinkers and den Dulk 1999; Takahashi 2007). Others find that a fraction of a population may be inclined to act with non-reciprocal altruism, even if the majority chooses otherwise (Johannesson and Persson 2000). Deep altruism similarly features an extreme degree of non-reciprocity, with no direct benefits to the originator—who may even be long forgotten by the future recipients of the completed effort. Musk and Bezos represent wealthy individuals who aspire to leverage their resources toward bold ambitions that they will not personally see to conclusion. Perhaps they hope to secure their names in history through such grand investments, but possibly they are also motivated by the desire to improve human civilization's capabilities of intentionally cooperating across intergenerational and deep time. A benefactor acting out a sense of deep altruism requires a vision for the species that extends beyond their own life; such a person may be motivated by the desire to alleviate suffering in the world, increase the sustainability of civilization, or pursue other global objectives that remain beyond the capability of any individual or generation to solve. Altruism in general, and deep altruism in particular, may be a uniquely human response to the problems elicited by civilization itself. Selfless action for the well-being of others across deep time might be one of the only available approaches for building a better long-term future for humanity.

8.6 Conclusion

Deep altruism remains a viable option for supporting human ambitions to settle space, as long as the initial benefactor can effectively transition the management, leadership, and vision of the effort to subsequent generations. Deep altruistic projects have a duration that is effectively infinite, as the succession of operations across generations is more critical to success than estimating a completion date. Any effort based upon a deep altruistic model must effectively communicate the vision that inspired the founder, and likewise must establish a secure source of funds that can also persist across generations. Wealthy individuals and institutions could conceivably finance such ventures, as long as they consider the benefits to the distant future as an adequate justification for investment today.

From the perspective of space settlement, which necessarily must operate with deep time in mind, the antithesis to deep altruism can be approximated as "deep egoism." A mindset fostered by deep egoism would assess the value of long-term investment by its propensity to benefit self, kin, colleagues, and descendants over others. The calculated return on investment for some asteroid mining ventures is predicted to exceed trillion of dollars; although these profits may not be realized by today's investors in asteroid mining technology, this conventional funding model would shunt the resulting wealth to the hands of individuals or organizations intended by the first investors. A long-term effort at extracting space resources for the purpose of building corporate or government wealth could be another approach to improve humanity's ability to operate over long timescales. It is important to note that deep egoism could conceivably sustain long-term efforts such as space settlement. For-profit entities face their own internal or external pressures to continue sustaining operations as long as they remain profitable; such pressures can drive innovation and allow companies to adapt to changing market pressures and new technology. Deep egoism could even motivate efforts across deep time and will likely be a significant driver of the near-term human exploration of space. Deep egoism resonates more strongly with modern capitalist ideals, although such an approach would risk failure if it is unable to continually provide a return on investment.

Humanity can take steps toward enabling altruistic efforts over deep timescales. Government and private granting agencies could develop competitive funding programs with decadal and longer performance periods in order to promote intergenerational pursuits of informational value. Wealthy individuals and organizations also hold the resources to establish their own long-term efforts by planning for intergenerational succession from the start. Crowdsourcing provides yet another approach for the general public to engage in supporting long-term efforts, with tools such as online crowdfunding and distributed consensus decision-making models providing the foundations for crowd-driven deep altruism. Striving toward informational value across intergenerational timescales will pave the way for extending such pursuits into deep time.

Space settlement will unfold in a piecewise manner, likely by a combination of government and private actors with a range of motivations. Commercial interests remain an important driver of the near-term space economy and could be a significant factor in developing the physical infrastructure required for space settlement. But deep egoism alone may be insufficient to sustain space settlement across millennia. If humanity genuinely intends to develop permanent settlements on Mars and in space, then it will inevitably be forced to develop new institutional governance models driven by deep altruism.

Acknowledgements I thank Lauren Seyler, Sanjoy Som, and Martin Elvis for helpful suggestions. This research did not receive any specific grant from funding agencies in the public, commercial, or not-for-profit sectors.

References

Brinkers, M. & den Dulk, P. (1999). The Evolution of non-reciprocal altruism. In D. Floreano, J. D. Nicoud, & F. Mondada (Eds.) *Advances in artificial life*. Springer.

Bruhns, S., & Haqq-Misra, J. (2016). A pragmatic approach to sovereignty on Mars. *Space Policy, 38*, 57–63.

Cockell, C. S. (2015). *Human governance beyond earth: Implications for freedom*. Springer.

Ehrenfreund, P., Race, M., & Labdon, D. (2013). Responsible space exploration and use: Balancing stakeholder interests. *New Space, 1*, 60–72.

Haqq-Misra, J. (2015). Toward a sustainable land use policy for Mars. In C. Cockell (Ed.) *Human governance beyond earth: Implications for freedom* (pp. 43–49). Springer.

Haqq-Misra, J. (2016). The transformative value of liberating Mars. *New Space, 4*, 64–67.

Haqq-Misra, J. (2018). Funding the search for extraterrestrial intelligence with a lottery bond. *Space Policy*,. https://doi.org/10.1016/j.spacepol.2017.03.003.

Jarvis, W. E. (2002). *Time capsules: A cultural history*. McFarland.

Johannesson, M., & Persson, B. (2000). Non-reciprocal altruism in dictator games. *Economics Letters, 69*, 137–142.

Musk, E. (2017). Making humans a multi-planetary species. *New Space, 5*, 46–61.

NASA. (2015). NASA's journey to Mars: Pioneering next steps in space exploration. Washington, DC, NP-2015-08-2018-HQ.

Seedhouse, E. (2010). *Martian outpost: The challenges of establishing a human settlement on Mars*. Springer Science & Business Media.

Takahashi, T. (2007). Non-reciprocal altruism may be attributable to hyperbolicity in social discounting function. *Medical Hypotheses, 68*, 184–187.

Trivers, R. L. (1971). The evolution of reciprocal altruism. *The Quarterly Review of Biology, 46*, 35–57.

Chapter 9
Interplanetary Sustainability: Mars as a Means of a Long-Term Sustainable Development of Humankind in the Solar System?

Andreas Losch

Abstract The sustainable development of Earth is a topic high on the agenda of the UN and any nation that cares for humankind's survival on its home planet. Economic, ecological, social, and cultural dimensions of sustainability are already under discussion. Taken the long-term development of the solar system into account, it is, however, unavoidable to include in these deliberations not only a technological imperative but also an expansion into space, because one distant day, the sun will become too hot to sustain life on Earth. Besides, asteroid impact could reshuffle evolution on Earth. Following NASA's initiative, and building on the UN'S considerations, a truly planetary sustainability therefore has to take the space surrounding Earth into account, and even would have to be extended to an interplanetary concept of sustainability one day. Would Mars be the planet to go, if sufficient bases will have been built or even terraforming efforts have taken place? How can we hold up the necessary dimensions of sustainability in this context? An outlook will discuss, which ethical questions would arise, if we would encounter extraterrestrial life while expanding into our solar system.

9.1 #Globalgoals Answering a Global Crisis

In 2015, the United Nations of planet Earth agreed on 17 sustainable development goals. They range from "No Poverty", "Zero Hunger", and "Wellbeing", passing societal issues like education and gender equality on to "decent work and economic growth." Environmental issues do not always come first; "climate action" is no. 13, for instance (https://sustainabledevelopment.un.org/sdgs). What is missing in any case is an 18. Sustainable Development Goal concerning the planet's space environment. This is the case, because humankind usually tends to forget that it lives on a limited

A. Losch (✉)
Faculty of Theology, University of Bern, Bern, Switzerland
e-mail: andreas.losch@theol.unibe.ch

Center for Theological Inquiry, Princeton University, Priceton, NJ, USA

K. Szocik (ed.), *The Human Factor in a Mission to Mars*,
Space and Society, https://doi.org/10.1007/978-3-030-02059-0_9

planet in vast space, even when there is currently much happening in the near-Earth environment.

The sustainability debate is a heritage of the ecological crisis that started to be realized in the 1960s. "No witchcraft, no enemy action had silenced the rebirth of new life in this stricken world. The people had done it themselves" (Carson 1962), as biologist Rachel Carson put it in her *Silent Spring* 1962, which started the environmental debate. Aurelio Peccei put it this way: "The world is already living in a state of emergency. But nobody wants to believe this. I am not conjuring up ghosts of the future but a state that already exists." (quoted in Amery 1972, p. 10) While the authors of the report of the Club of Rome voted for an equilibrium model as countermeasure to the crisis, the UN sustainable development goals do include "economic growth" (no. 8). Out of our dealing with the ecological crisis, the idea of sustainable development evolved, which distinguishes itself from previous analyses in acknowledging this *need* for economic development.[1]

The term "sustainability" was first prominently used by the so-called Brundtland commission in 1987. According to this commission's official definition, which shall be used here as well, sustainable development is a "development that meets the needs of the present without compromising the ability of future generations to meet their own needs," while these needs are "in particular the essential needs of the world's poor" (World Commission on Environment and Development 1987, pp. 16.41). Sustainability in political perspective therefore is a concept that includes besides the economic–ecological dimensions also an important social dimension from the start.

This eco-social perspective may be a heritage of the early discussion of the theme in context of the Ecumenical Council of Churches. Already in 1974, therefore, more than a decade before the Brundtland commission, in Bucharest an ecumenical world conference on "Science and Technology for Human Development" took place. At this conference, the term "sustainable and just society" was first used (Lienemann 2007, p. 101). From here we can understand why there is a social dimension of justice within the concept of sustainability according to the Brundtland Commission. Also, according to Pope Francis' encyclical letter *Laudato si* we have to acknowledge, "that a truly ecological approach *always* becomes a social approach; it must integrate questions of justice in debates on the environment, so as to hear *both the cry of the earth and the cry of the poor*." (Francis 2015, p. 49)

This is true in an even deeper sense than it may seem at first hand. On the one hand, the concept of sustainability introduces the idea of *intergenerational justice*. The poor of the future will be the first to suffer from a future lack of resources. On the other hand, the problem with the sustainability concept is that the poor of the present must pay for it. In industrial societies, sustainable development involves necessary structural change where exactly those that have already been exploited in the past must carry the more societal costs of the changes. The change *managers* get well paid, and wealthy people can easily pay for fair traded and more ecological food and energy. Coal miners, however, lose their job, and fishers do not understand why they need to stop expanding their small profit margin with catching more fish. Of course,

[1]The following two paragraphs are taken from Losch (2018, p. 3).

if they go on like that, there will be no fish any more to catch, but that is not easy to foresee for a fisherman. At least in my view, from "Trumpamerica" to "Brexit" this *intragenerational* conflict surrounding the sustainability concept marks the political developments of our days and it is mirrored in the composition of the political parties and the interest they represent. This is a highly important conceptual challenge and needs to be reflected on much more in depth than it can be done here.

9.2 Planetary and Interplanetary Sustainability

What I call "planetary sustainability" (Losch 2018) I want to define as a *concept of sustainability that is aware of the fact that Earth is a planet*. For long, humankind knew and regarded Earth as a *globe*, this knowledge was even part of the Aristotelian worldview (cf. Russell 1991). The news of our Age is, hence, less the actualization of this knowledge than the truly global scope of economic competition and cooperation. Nevertheless, we still have collectively to become aware of the fact that we live on a planet with limited resources (Rockström et al. 2009; Steffen et al. 2015), and with a space environment.

In this spirit, NASA launched in 2014 an initiative with the title "planetary sustainability", whose vision statement includes three main objectives:

1. A world in which all people have access to abundant water, food, and energy, as well as protection from severe storms and climate change impacts;
2. Healthy and sustainable worldwide economic growth from renewable products and resources;
3. A multi-planetary society, where the resources of the Solar System are available to the people of Earth.

If we apply this planetary perspective on the question of sustainability, a new dimension emerges: later generations should be able to meet their own needs without perishing due to events in our solar system, and hence be able to fulfill at least their most substantial need, which is survival.

The idea of planetary sustainability does combine two statements: the "development that meets the needs of the present without compromising the ability of future generations to meet their own needs," (World Commission on Environment and Development 1987, p. 16) and the fact that "without our expansion of our instruments and people into space, humanity could conceivably perish." (Pass et al. 2006, p. 5) Hence, acknowledgement of the planetary dimension of the sustainability questions leads automatically to the issue of interplanetary sustainability: where should, where could humankind survive? Events in our solar system that could endanger our survival include not only our sun turning into a red giant and hence an environment-caused heat death (if we can avoid the Anthropocene's (Steffen et al. 2014) human-caused heath death at all), but also the chances for an asteroid impact reshuffling evolution on planet Earth like what happened to the dinosaurs.

9.3 Reflection: Is the Sustainable Survival of Humankind of Value?

Are we hence responsible to develop techniques, which allow our late offspring generations to survive somewhere else, to colonize other planets so that they can at least meet their most basic need, which is survival? If so, we need to embrace and steer the technical developments, and are not allowed to retreat into an Eco-romanticist cave. What might be an individual option is no option for humankind as a whole.

Depending on the stance taken regarding environmental ethics, the answer to the question of this section may vary. Would the planet be better off without humans? Could another asteroid impact actually do good in allowing evolution to reconsider its trajectory? Let us shortly consider the assumptions of the individual ethical stances in due course (cf. Persson 2012).

(a) *Anthropocentrism*: what is regarded in some circles, including the UN, as a virtue, might be a problem or even a vice as well. What allows us to put "humanity in the forefront" (UNESCO 2003) and assigning moral status only to ourselves? Regarding lesser aims (than its survival) of humankind itself putting our deepest concerns on top might be a reasonable choice, but what actually makes us believe that humankind does have a special status? Theologically (in the Jewish-Christian tradition), this idea is quite clearly connected to the idea of being created in the image of the Godhead. But for those who do not live in this theological tradition: are we allowed to put ourselves first, just because it is us who makes these deliberations? Does it help to claim that a particular property is exclusively human? Couldn't there be other rational species in the universe or even on our planet, so at least a more universal *ratiocentrism* would be advisable?

(b) *Sentientism*: a response to this challenge of human uniqueness might be to focus on feelings like pleasure and pain as a feature that would include the highest moral status as well. Any animal that can feel would then have to be regarded as somewhat "sacred", and we would need to consider sharing the planet with these creatures for moral reasons alone. What comes as a humble and modest approach on the one hand could endanger our survival on the other hand: if the bear feels hungry, he is allowed to prey on that child in the near and should not be shot—if this would provide a valid scenario? Why should the human species be the only species to limit its powers in such a way?

(c) *Biocentrism*: an even more general approach could be to grant moral status to any living being. This can be tricky regarding microbes that make us sick or actually any source of food we use to nourish. A stance like this would lead us in any case very quickly to the question "What is Life?", which is still not completely answered (Losch 2017), are there actually living beings or is life itself a holistic phenomenon. We should at least be aware of the fact that life forms are entangled on a multitude of layers, mutually depending on other life forms, so, in my regards at least, this stance does not differ much from the next one, which is holism.

(d) *Holism/Ecocentrism*: when we experienced the "overview effect" the first time, recognizing Earth as a whole, a blue marble at the horizon in the endless cosmic void, we realized that the planetary systems are naturally intensively interconnected. Any appreciation for life of whatever kind on this planet must logically include an appreciation of its living environment as well, because else life on planet Earth could not be sustained. One can, however, also be an ecocentrist counting lifeless celestial bodies as having a moral status of their own (Losch 2019). Yet what about the role of the individual? Individuated dignity is a stern moral principle of our days, stemming from Immanuel Kant's considerations. This does not seem to be compatible with a holist (or ecocentrist) approach.

My conclusion for these ethical dilemmas is a universal *ratiocentrist* point of view (cf. Smith 2009), while being open which species would have to be included in this perspective. Without mind, there wouldn't be values at all; hence, we need to focus on mindful or rational species to allow for *any* sorts of ethics. Two possible objections to this claim (regarding the dependency of values on mind) are (1) a Mastermind behind the universe (e.g., Polkinghorne 1998) or (2) assuming a Platonic noetic realm. But without *believers* in this Mastermind or *apprehenders* of the Platonic realm, the values would be there, but not being effective. In my point of view, mind can only exist in this world based on matter, so it needs an animal's body, a living environment, and a stable planetary system to survive in the long run.[2] Hence, all stances have their moment of truth in them. That's why I do not disregard the foci of the other stances, but regard them as necessarily *complementary* point of views (within a ratiocentrist setting).

A fundamental question here to be evaluated is, what is a (living) system? What and where are the fuzzy borders of live and non-live? Are there any limits, or how deep are we connected with our environment, where do feelings belong and what is mind? All these questions lie behind these ethical distinctions and I fear in our time it will be almost impossible to find an answer here (Losch 2019).

9.4 Survival on Mars: Terraforming the Planet or not?

If we need to move into space to survive in the long run, one should welcome recent space mining efforts and evaluations of such enterprises as our beginnings to make that move into space. Of course, there are many questions to be answered and issues to be solved, how such space mining can be considered and performed in a sustainable way, especially when it comes to the social aspect of sustainability. Luxembourg with

[2]If there would really exist one day a GAI (general artificial intelligence) that would have the capability for self-reproduction, also other environments than our known one might prove making sense in these regards.

its *spaceresources.lu* initiative is currently in the lead in these regards and committed to a sustainable approach, which nevertheless still has to be worked out and clarified.

The next place to move one day could be Mars. First of all, there is the question, whether Mars is lifeless or not (Cockell 2014; cf. Losch 2017, p. 1). Here, we face similar dilemmas like depending on our stance regarding environmental ethics on Earth. Most probably there could only be "extraterrestrial" *microbes* on Mars, most likely deep under its surface (Ivarsson and Lindgren 2010), so only *biocentrists* or *holists* would raise objections against a potential "terraforming" treatment of Mars, if one day possible. As we, however, have argued that all ethical stances have their right to be considered and are even necessary as complementary approaches to a universal ratiocentrist point of view, we must consider the potential issue of life on Mars very seriously.

Even if we do not grant microbes moral status, a multitude of necessary ethical considerations remain. ET microbes might prove of some sorts of value to us in any case and hence would need to be protected. This might consist in any of the instrumental values known to us, for scientific/epistemic, aesthetic, monetary, or historic reasons. Or maybe we have to grant any form of life an *end* value in itself, while we restrict moral status to rational beings? What may sound just at first sight is problematic, as in fact we always have to do an evaluation when sending a probe to Mars. This is required by the so-called planetary protection programs (Losch 2018, p. 4–5). When we sterilize the probe, we kill the (most of the) Earth microbes on it; only after this process, it can actually serve its function to detect ET life on Mars without just reporting Earth microbes it brought itself to the red planet.

Now, if there would be microbial life on Mars: would it be allowed to "sacrifice" it when "terraforming" Mars one day? If humans can warm the Earth beyond measure, why not take care of frozen Mars (McKay 2009)? Due to its small size (Mars is 1/10 the mass of Earth), several important factors are divergent concerning the habitability for life. In sum, however, "the necessary material to construct a biosphere are likely to be present on Mars and … the fundamental physical aspects of Mars that would be virtually impossible to alter … are similar to corresponding values on Earth." (McKay 2009, p. 251) The one exception to that is the low surface gravity, but life could hopefully adapt to that.

The search for life on Mars is out because of the remnants of liquid water that have been observed on the planet, so Mars is a planet that once has been more habitable. In general, a restored habitability of Mars would only last about 10–100 million years. Given the fact that the remaining life span of Earth is about 500 million years only (the heating of the Sun will make it Venus-like) (Caldeira and Kasting 1992), an approach in these regards nevertheless makes sense. "No solution for infinite lifetime exist for Mars or Earth" (McKay 2009, p. 249), so also an ecosynthesis—as McKay calls "terraforming" of Mars—can only present a provisionary means. The question remains how much time Mars would buy us or if we shouldn't plan for longer distant travel into space from the scratch. In any case, it would certainly prove a splendid test case (cf. McKay 2009, p. 256), for settling or for more fundamental approaches to dead planets in the habitable zone to their star.

McKay also attempts at an answer to the ethical question: should we really try to terraform Mars? He observes three principles behind systems of environmental ethics, which are emphasized in different degrees.

1. Preservationism—the fundamental principle that nature is not to be altered by human beings.
2. Wise stewardship—the fundamental principle that the measure of all things is utility to humans, in the broadest and wises sense of utility.
3. Intrinsic worth—the fundamental principle that there exist sets of objects which have intrinsic worth regardless of their instrumental value to humans (McKay 2009, p. 255).

The third principle is the least developed, as McKay admits. He, however, wants to employ this principle in a way that suits an ecosynthesis of Mars: "If richness and diversity in life forms is a value in itself, then planetary ecosynthesis on Mars is a good thing." (McKay 2009, S. 256) I must answer this claim with two reservations: (1) this is only the case if there really is no whatsoever life existing on present Mars (Persson 2014), else an ecosynthesis would equal a gargantuan extinction event. I would differ from McKay in the regards that even if life on Earth has its origin on Mars, the independent development of the Martian "biosphere" would need to be carefully tracked. To state that we could risk existing life on Mars "if we do not alter the Martian environment so as to enhance its global habitability" (McKay 2009, S. 258) is a bit crude disregarding the intrinsic interconnection between life and its environment in my regards. (2) McKay misses to discuss aspects of intrinsic worth attributed to (more or less) "untouched" landscapes. Especially, a sustainability account emphasizing a necessary cultural dimension would be able to embrace planetary parks (Cockell and Horneck 2004) as "world heritage sites" (Losch 2019).

So I would go along with ratiocentrist considerations, maybe adding that Smith's emphasis on long-term (rather than short-sighted) ratiocentrism, which is aware of the environmental interconnections of rational life (Smith 2009), could be dubbed "sustainable ratiocentrism" or such.

9.5 Conclusion: Interplanetary Sustainability

A large-scale assessment of the question of sustainability leads necessarily to the issue of interplanetary sustainability. NASA is to be congratulated by having made the connection early. What has been discussed here regarding Mars has to be considered regarding other potentially habitable locations in our solar system and beyond as well (e.g., the moon *Europa*). Also on the level of the United Nations, the Space2030 initiative attempts at least at adding sustainability considerations on near-Earth space. An 18th Sustainable Development Goal focusing on our space environment would do good in this regard, but as politics takes time, it could probably only be considered in the next iteration of UN goals. Of course, no such considerations should distract us from keeping our current focus on the one home we currently have, which is planet

Earth. That is why the main thrust of the Space2030 initiative is how space science and space devices can contribute to the sustainability of Earth. The cultural dimension of sustainability is still fuzzy. There are some aspects of sustainability (like world heritage sites or planetary parks) which could be missed without regarding culture as a sustainability pillar of its own; as a counter-term to be distinguished from "nature", culture is a very broad term in more cautious regards.

9.6 Outlook: Extraterrestrial Encounters

The issue of the potential existence of extraterrestrial life could alter the considerations undertaken here substantially. What has been discussed regarding ET microbes on Mars (or likewise in Europa's subsurface oceans) might take an entirely different turn when applied on extraterrestrial *intelligent* life. The question of course is, whether (and under which conditions) any life has the chance to develop intelligence.[3] If it remains a unique phenomenon on Earth, the considerations regarding the necessity of humankind's survival may hold. When acknowledging the possibility of ETI in the universe, however, a universal and sustainable ratiocentrist account must be modified.

My claim that "without mind, there wouldn't be values at all, hence we need to focus on mindful or rational species to allow for *any* sorts of ethics," does not affect the question whether the survival of humankind is *necessary* or not. ETI provided, humankind's survival certainly still is of value, as humankind as a species involves mind and values; if there are other rational species with other values, however, having moral status as well, there could be a decisive clash of cultures taking place. That is the case, unless the *is/ought division* proves only preliminary valid and unless we know more about the nature of competition and cooperation in evolution one day, which would help closing the division. Else, we could face a very particular problem: how to evaluate fundamental differences regarding our values.

As best way to deal with such a situation, probably the way we can already now train on Earth would prove valid: talking and listening to others and find out what we share in our ethics and cultures, like, e.g., the golden rule (Küng 1990). Could it be possible that this rule is a rule built into evolution itself, that is, could it be part of nature?

References

Amery, C. (1972). *Das Ende der Vorsehung: Die gnadenlosen Folgen des Christentums*. Reinbek bei Hamburg: Rowohlt.

Caldeira, K., & Kasting, J. F. (1992). The life span of the biosphere revisited. *Nature, 360*(6406), 721–723. https://doi.org/10.1038/360721a0.

[3]For a gradual account of the intelligence concept, see Peters (2017).

Carson, R. (1962). *Silent spring*. Boston, Cambridge: Houghton Mifflin Company, The Riverside Press.
Cockell, C., & Horneck, G. (2004). A planetary park system for Mars. *Space Policy, 20*(4), 291–295. https://doi.org/10.1016/j.spacepol.2004.08.003.
Cockell, C. S. (2014). Trajectories of martian habitability. *Astrobiology, 14*(2), 182–203. https://doi.org/10.1089/ast.2013.1106.
Francis. (2015). Laudato si: On care for our common home. Retrieved June 8, 2018, from http://w2.vatican.va/content/francesco/en/encyclicals/documents/papa-francesco_20150524_enciclica-laudato-si.html.
Ivarsson, M., & Lindgren, P. (2010). The search for sustainable subsurface habitats on Mars, and the sampling of impact ejecta. *Sustainability, 2*(7), 1969–1990. https://doi.org/10.3390/su2071969.
Küng, H. (1990). *Projekt Weltethos*. München: R. Piper.
Lienemann, W. (2007). «Sustainability » in Ökumene und Theologien. In Ruth. Kaufmann, Paul. Burger, & Martine. Stoffel (Eds.), *Nachhaltigkeitsforschung–Perspektiven der Sozial–und Geisteswissenschaften: Recherche dans le domaine du développement durable–perspectives des sciences sociales et humaines* (pp. 99–122). Bern: Schweizerische Akademie der Geistes–und Sozialwissenschaften.
Losch, A. (2019). Challenges of cultural sustainability on an (inter)planetary scale. In T. Meireis & G. Rippl (Eds.) *Cultural dimensions of sustainability: Perspectives from the humanities and social sciences*. Routledge.
Losch, A. (Ed.). (2017). *What is life? On earth and beyond*. Cambridge, UK: Cambridge University Press.
Losch, A. (2018). The need of an ethics of planetary sustainability. *International Journal of Astrobiology* 1–8. https://doi.org/10.1017/s1473550417000490.
McKay, C. P. (2009). Planetary ecosynthesis on Mars: Restoration ecology and environmental ethics. In C. M. Bertka (Ed.), *Exploring the origin, extent, and future of life: Philosophical, ethical, and theological perspectives* (pp. 245–260). Cambridge, UK, New York: Cambridge University Press.
Pass, J., Dudley-Rowley, M., Gangale, T. (2006). The cultural imperative to colonize space: An astrosociological perspective. *Space 2006*. San Jose, California. [Reston, Va.]: [American Institute of Aeronautics and Astronautics]. https://doi.org/10.2514/6.2006-7488.
Persson, E. (2012). The moral status of extraterrestrial life. *Astrobiology, 12*(10), 976–984. https://doi.org/10.1089/ast.2011.0787.
Persson, E. (2014). What does it take to establish that a world is uninhabited prior to exploitation? A question of ethics as well as science. *Challenges, 5*, 224–238.
Peters, T. (2017). Where there's life there is intelligence. In A. Losch (Ed.), *What is life? On earth and beyond* (pp. 236–259). Cambridge, UK: Cambridge University Press.
Polkinghorne, J. C. (1998). *Belief in god in an age of science*. New Haven, Conn., London: Yale University Press.
Rockström, J., Steffen, W., Noone, K., Persson, Å., Chapin III, F. S., Lambin, E., et al. (2009). Planetary boundaries: Exploring the safe operating space for humanity. *Ecology and Society* 14 (2). https://doi.org/10.5751/es-03180-140232.
Russell, J. B. (1991). *Inventing the flat earth: Columbus and modern historians*. New York: Praeger.
Smith, K. C. (2009). The trouble with intrinsic value: an ethical primer for astrobiology. In Constance. M. Bertka (Ed.), *Exploring the origin, extent, and future of life: Philosophical, ethical, and theological perspectives* (pp. 261–280). Cambridge, UK, New York: Cambridge University Press.
Steffen, W., Broadgate, W., Deutsch, L., Gaffney, O., & Ludwig, C. (2014). The trajectory of the anthropocene: The great acceleration. *The Anthropocene Review, 2*(1), 81–98. https://doi.org/10.1177/2053019614564785.
Steffen, W., Richardson, K., Rockström, J., Cornell, S. E., Fetzer, I., Bennett, E. M., et al. (2015). Sustainability. Planetary boundaries: Guiding human development on a changing planet. *Science (New York, N.Y.)* 347(6223), 1259855. https://doi.org/10.1126/science.1259855.

UNESCO. (2003). Planetary sustainability in the age of the information and knowledge society for a sustainable world and future, Paris.
World Commission on Environment and Development. (1987). *Our common future.* Oxford u.a.: Oxford University Press.

Chapter 10
Mars Environmental Protection: An Application of the 1/8 Principle

Tony Milligan and Martin Elvis

Abstract There are a number of candidate rationales for the settlement of Mars. These are considered in Sect. 10.1. At least one of them is economically plausible: its use as a base of operations for asteroid mining in the Main Belt. This rationale suggests that environmental protection on Mars needs to be considered in a broader context than that of the planet alone. More specifically, the authors argue in Sect. 10.2 that planetary environmental protection (which goes beyond 'planetary protection' in the stricter legal sense) is partly a matter of containment and so requires a framing principle. Section 10.3 supplies such a principle for protection, "the 1/8 principle": while economic growth in space remains exponential and comparable to historical patterns on Earth, we should use no more than 1/8 of the available resources, in order to secure a reasonable 'breaking distance' before we reach a point of super-exploitation. Section 10.4 examines the difficulties of applying such a principle to Mars, the need for trade-offs and special weighting for special strategic resources such as uncollapsed lava tubes. Section 10.5 points out the ease with which the principle can be turned into policy options and its good matchup with human practices of valuing and use. The latter considerations may not be decisive, but they do count strongly in its favour.

10.1 The Rationale for Settlement

Mars occupies a unique place in terms of human use of the solar system. There are two reasons for this: surface area and location. It is within reach, conveniently situated for expanding the human activity to the Main Belt asteroids, and accounts for a large portion of the available surface area that we can work with (for whatever purposes).

T. Milligan
Department of Theology and Religious Studies, King's College London, London, UK

M. Elvis (✉)
Harvard-Smithsonian Centre for Astrophysics, Cambridge, MA, USA
e-mail: melvis@cfa.harvard.edu

K. Szocik (ed.), *The Human Factor in a Mission to Mars*,
Space and Society, https://doi.org/10.1007/978-3-030-02059-0_10

10.1.1 Surface

It is by no means certain that humans will ever go anywhere beyond the Solar System. The sheer immensity of interstellar space may pose an obstacle too great to overcome. What exists here, inside the Solar System, may be all that we (and our successors) will ultimately be able to work with. Yet, we cannot use most of what is *here*, but only a fraction of it. After all, the Sun accounts for most of 'what exists here', taking up more than 99.8% of the system's total mass, leaving us less than 0.2% of the total mass to work with. When we think of matters in terms of surface area the prospect does not improve. Setting aside the Sun, the gas giants (Jupiter, Saturn, Uranus and Neptune) account for over 98% of the remaining surface area. This is a problem. While they are some distance away, we might still reach them without difficult-to-imagine technologies. Not now, but someday. Leaving their surfaces would, however, be much harder. Their mass is immense and their gravity wells are deep. The energy required to climb out of them would be of a similar scale. In the cases of Neptune and Uranus, this might be possible (again, not now, but in the future) but Saturn and Jupiter look prohibitively difficult. The idea of their commercial use, even using automated systems rather than flesh and blood creatures, is problematic. Their integration into any manner of multi-planetary system of human settlements seems unlikely or, at least, open to question. Minimally, we cannot safely assume that this will happen, even at some unknown point in the future. We cannot automatically factor the availability of these places into our assumptions.

Accessible surface, of significant scale, may well come down to the 2% that remains once we discount the gas giants, the surfaces of the inner planets, some of the larger asteroids (such as Ceres and Vesta) and various moons near and far. Of the latter, there is our own Moon and there are the Galilean moons of Jupiter, large enough to have been spotted with seventeenth-century telescopic equipment: Io, Ganymede and Callisto (which are slightly larger), and Europa (which is slightly smaller). There is Saturn's moon Titan (larger) and Neptune's moon Triton (smaller) but averaging out in the same broad territory as the Moon. In addition, there are the dwarf planets and various other bodies, all of which are significantly smaller. The moons themselves are all significantly smaller than the Earth, Mercury, Venus and Mars. In fact, with the exception of Venus, every accessible surface turns out to be much smaller than the Earth. We take up a quarter of the 2% (An additional reason to regard the Earth as truly remarkable and irreplaceable).

Given their weaker gravity, and the reduced energy requirements for leaving them (again excepting Venus), it is tempting to think of all these worlds as potential sites for a human presence. However, Io could be too exciting with its volcanoes; Titan has a poisonous atmosphere; Europa has a thick icy crust. All these worlds may be ours, but the 'all' is less than we might once have believed. The bottom line is that there are four good candidate moons in the solar system (the Moon, Ganymede, Callisto and Triton). Mercury and Venus are hellishly Hot. Eventually, advanced terraforming techniques may make more worlds habitable, but that is too distant a prospect to concern us now. After the Earth, that leaves just one other unusually suitable planet, i.e. Mars. The surface of Mars turns out to be too precious. Of course, it is not *only* a resource. The solar system is not merely a giant quarry. But use-as-resource is one aspect of our relation to it, and crucially, it is a *limited* potential resources.

10.1.2 Expansion

The background assumption here is, of course, that there is a good rationale for humanity to make use of planetary, lunar, martian and other surfaces, i.e. a rationale for use to go *somewhere*. This is an assumption which is notoriously difficult to justify in simple terms, or indeed in any terms at all (Schwartz 2018). There seems to be no single clear-cut justification that silences or obviously outweighs all conceivable objections. Even so, many people do seem to have a strong intuition that space settlement is not just ethically defensible but would constitute progress for humanity, or for life itself. Widespread and sustained intuitions are not always defensible, but neither are they easily dismissible. They have a place in initial deliberations even if, at some later point, they may be set aside. This is a familiar methodological point within ethical theory, one associated with the idea of pursuing a ‘reflective equilibrium’ in which intuitions are allowed to shape overall sets of principles and theories, which then lead us to re-evaluate the plausibility of the particular intuitions (Rawls 1974). Methodologically, something of this sort, with allowances made for qualification and nuances, may be ‘the best game in town’.

In line with this methodological assumption, we may say the following: the idea that there may well be a duty to extend our presence, or the presence of life, is a particularly difficult intuition to set aside. This does not mean we have to ‘be fruitful and multiply’ (Genesis 1:28) unthinkingly but the strength and widespread nature of the intuition may well support the case for regarding it as the initial default option. Familiar attempts to claim something *more* than this have, of course, been tried. Appeals to a variety of different considerations: it is better to be ‘open’ to expansion than ‘closed’ to what is other (Turner 2015); the idea that space settlement can improve our attitude towards the Earth; the value of science which requires that we expand our presence; the importance of backing up our terrestrial biosphere; human expansion as a vector for life in general and, in the absence of any certainly about the existence of life elsewhere, we may have a duty to expand life; space offers prospects for a Tsiolkovsky-style utopia, or perhaps something a little more modern but equally utopian (Tsiolkovsky 2004). Appeals have even been made to a special kind of political rationale, to the prospect that space offers as a continuation of our liberation from the aristocratic legacy of Europe and that we are economically committed to indefinite expansion rather than to any more stagnant domain (Zubrin 2012).

In the more questionable column, we should probably place appeals to anything like ‘Manifest Destiny’ or ‘taking what is ours’ with their colonialist overtones, yet such ideas have their supporters (Green 2017). This is not, however, the nineteenth century. Progress (whatever it may be) need not be dominion. Actual colonial projects strike most of us as morally flawed (for good reasons). Our focus is on the basic economic point that the Solar System will not permit indefinite expansion. It is not large enough for this to occur or (as we shall see below) large enough for the dynamics of expansion to be unimportant.

Focusing upon whichever cluster of the above considerations we regard as most plausible, it is tempting to say that while none of these reasons seem obviously compelling, but 'look at the number of them'. In combination, they align and *may* possibly add up to something a little more compelling, like a bundle of sticks which are strong enough to support a load when joined together. But the jury is still out. Certain kinds of weak justifications still remain weak in combination. Perhaps, a further argument is required about how these reasons might combine. Nonetheless, they certainly seem to match well with the methodologically driven assumption that some extension of human presence should be regarded as the default option. And this assumption, in turn, aligns well with the sheer pragmatics of space. Whether we like it or not, humanity *is* increasing its off-world presence, and it *is* likely that this will continue. While this expansion could remain restricted to a near-Earth economy, because of the length of time that larger scale investments would take to yield a return, there is at least one clear rationale for thinking that it will not do so and that broader expansion is the likeliest prospect.

10.1.3 Location

The rationale is straightforward. The utilisation of space resources on *any* significant scale will require space mining, and particularly the mining of metallic asteroids. Inconveniently, the number of asteroids that are of the right size and composition and to which we can send suitably massive mining equipment, given our present rocketry, is surprisingly small. Maybe as few as a dozen today but growing to hundreds as we find more and more powerful rockets are developed (Elvis 2014). The SpaceX Falcon Heavy or its prospective replacement, the Falcon Super Heavy, or Blue Origin's New Glenn and/or New Armstrong may bring us closer to this prospect. The obvious solutions to the problem of access are that we should try to shift more asteroids towards us (technically feasible, given foreseeable future technologies, but difficult and also somewhat dangerous) or that *we* must go to *them*. Again, the latter looks like the default. Sustained and large-scale asteroid mining may well require that we extend our human presence out at least as far as the Main Belt asteroids. And that is where Mars enters the picture. If we are ever seriously in a position to engage in Main Belt mining, then we will need an initial base of operations from which to do so. Preferably, one with a good orbit, and also with a far shallower gravity well than the Earth. Mars looks like the obvious candidate. A mining base on Phobos would have tens of thousands of sizable Main Belt Asteroids as accessible as the most easily reached NEAs from low Earth orbit (Taylor et al. 2019). Mars looks like the only plausible candidate. We could, of course, revive the idea of a mobile off-world habitat, an O'Neill cylinder of some kind (O'Neill 2001). However, quite apart from the safety concerns that this would raise, such a large project would only be viable given existing and extensive asteroid mining operations. It would be a case of putting the vehicle (possibly some analogue of the Titanic, if we move too quickly) before the horse.

What follows from these rough and ready considerations is surprisingly clear. Even if there were no resources of worth on Mars, the pragmatic rationale for establishing some sort of stable human infrastructure and presence looks strong. Mars settlement (of some scale) and extension of mining to the Main Belt asteroids may well have to go together.

10.2 Protection as a Problem of Containment

The genuineness of this pragmatic rationale for establishing a presence on Mars drives us to consider environmental protection for Mars. 'Environmental protection' is not, however, just 'planetary protection' in the current and restricted sense used in space law and policy formation. The latter is presently limited to considerations of avoiding back contamination (returning anything dangerous to Earth) and limiting forward contamination (thereby spoiling sites of scientific interest by e.g. accidentally bringing our microbes to them). We will instead refer to 'planetary environmental protection' in order to stress our broader concerns.

Planetary environmental protection, as understood for the Earth, involves a range of safeguards for diversity and for unique objects and places so that current and future generations may enjoy and identify with them. Often it goes beyond human concerns alone or, at least, blurs the boundaries between human concerns and concerns that reach beyond the significance of the human. To some extent, this happens with appeals to the *integrity* of places and things (Milligan 2018). What we will point out, for now is that, insofar as the arguments for such terrestrial environmental protection draw upon considerations of uniqueness or irreplaceability, variants will also apply to Mars. This will hold irrespective of any further backstory that is then told about what we mean when we refer to the 'value' of anything or our reasons for 'valuing' anything, unless the backstory happens to deliberately build in special and exclusive claims about our relationship to the Earth. We have no intention of thinning our argument by adding unnecessary premises. When reasons for protection appeal are to uniqueness or irreplaceability *simpliciter*, such appeals will carry over and also apply to Mars.

Mars is, after all, a unique place with a distinctive history. It is not another Earth. The Martian terrain, Olympus Mons, the Valles Marineris and all of the features, which go to make it so distinctive, are plausible candidates for safeguarding against destructive use. This uniqueness does not, however, require us to take a hands-off attitude. It does, we argue, mandate that we consider environmental protection of Mars before we act.

One of the reasons why uniqueness does not require a hands-off attitude is that ethical deliberation is not exhausted by pointing out that there are reasons for doing x, or for not doing x. As a point about practical reason, there are often countervailing considerations: reasons which point in different sorts of directions and these are not easily silenced. This leaves moral agents such as ourselves with the difficult tasks of weighing up multiple interests (including those of *being good agents*, rival obligations and conflicting duties). These may sometimes have to be traded-off against one another, simplified and weighed in order to arrive at a decision under conditions of uncertainty. This is what a good deal of our actual experience of ethical agency is like. It is tempting to say that humans are naturally ethical pluralists, responsive to more than one thing at any given time and that ethical pluralism may be a beneficial legacy of our biological history. Alternatively, we can appeal to what goes deep within human socialisation in cultures such as ours (It is, by definition, an inbuilt feature of anything like a broadly liberal society). Any one-dimensional argument for unbridled settlement or hermetic protection is therefore liable to fail because it would require us to be people of a radically different sort. The diametrically opposed views that species expansion trumps all considerations, or the view that humans are a sort of virus from which worlds need to be protected, are both unrealistically monomaniacal. When it comes to ethics on an interplanetary scale, between worlds, our best response may have to turn upon an account of what good ethical agency and decision-making is like, because nothing is trumps.

In line with this approach, a plausible case for planetary environmental protection in the case of Mars needs not be based upon a denial that there *are* good reasons for wanting to go to Mars, for establishing a stable human presence there, and for using resources in situ. There may even (as suggested above) be a duty to go, or to take preparatory steps in that direction, although this is a noticeably stronger claim.

The issue of protection may, then, best be set in the context of an acceptance of both stakeholder interests and planetary integrity. By taking this approach, we can acknowledge the legitimacy of competing claims, including the legitimacy of at least some commercial claims of interest, while doing justice to the idea that it is worthwhile (and is, perhaps, an obligation) to protect the Martian-ness of Mars. And this remains the case even if we acknowledge the unavoidably transformational role of human landings, a sense that, in Erik Conway's terms, the Mars that scientists want to study will not exist anymore, but some other Mars will (Conway 2015). This approach does, however, also require that commercial interests are *not* viewed as automatically overriding. And this approach has a certain policy aptness. It is in line with the current wave of legislation from Europe and the US which favour the opening of pathways to commercial activity while endorsing the ongoing importance of other considerations (most obviously, those of science and of compliance with international law).

Irreversible changes and changes which might affect the integrity of the planet will, accordingly, have a special significance. As an extreme example, any attempt at melting the ice caps in order to thicken the Martian atmosphere would qualify. If we are ever in a position to attempt this, the project should not be evaluated in exclusively technological terms: *can* will not imply *ought*. We might then at least consider (as Kim Stanley Robinson does in his *Mars* trilogy) restricting changes to atmospheric density in order to preserving the original Martian-ness of Olympus

Mons and other high-altitude sites. These terraforming examples are, however, far beyond the immediate context of a more modest move towards mining operations in space.

As indicated above, it is our contention that the primary threat to what is worth protecting on Mars may well arise as a result of its strategic location. It may be thought of as a problem of containment in two senses. First, ensuring that broader space objectives do not *spillover* unduly onto the Martian surface (Milligan 2015). Second, constraining the *expansion dynamic* once surface activities do begin. Up to a point, this first of these considerations makes the protection of Mars contrast with lunar protection and may even put the two in tension.

For example, one of the Moon's special resources is Helium 3 (^{3}He), for fusion power generation. The prospect of asteroid mining for ^{3}He could relieve pressures upon the lunar surface in favour of processing the asteroid regolith (the loose surface material) that also contains it. While the density of ^{3}He is less on asteroids than on the Moon (due to the ten or more times lower solar wind flux in the Main Belt), we have fewer reasons for asteroid protection than for lunar protection, especially once the larger asteroids (for which claims of uniqueness and integrity might well be made) are set aside. ^{3}He mining, especially if we can reach the Main Belt could be a viable *alternative* to lunar mining. The Main Belt activity that might relieve pressures towards lunar mining may require at least some associated activity on Mars.

Thinking of matters from the standpoint of containment, both senses involve a significant shift. While the 'expansion dynamic' sense can be viewed from a strictly planetary attitude, the 'spillover' sense requires a shift to a broader context set by growth and development beyond Mars, placing it within the space economy *as a whole*. Given the limited number surface areas available to us, and the possibilities of irreversible changes on such surfaces, it makes sense to set planetary-level environmental protection within such a broader context. Approaches to planetary protection, even in our extended 'environmental protection' sense, have not always done this. Exceptions are the work on 'cosmocentric ethics' (e.g. Lupisella 2016), which may be problematic because it raises concerns about what a viable ethical theory has to look like and how broad our outlook can be, and the more accessible emphasis placed by Chris McKay (2013) upon diversity across the Solar System.

The difficulties of situating Martian protection within any kind of broader context are, of course, considerable. Given the limitations of available planetary and other surface for us to work with, it does, however, seem to be the right way to go. Surface is a limited resource. Our favoured pathway for doing so, set out in a more detailed case for treating an extensive portion of the Solar System as 'wilderness' (Elvis and Milligan 2019), involves an attempt to establish a large-scale constraint by establishing how much of the accessible Solar System we can safely use without running the risk of some reasonably proximate generation of humans facing resource exhaustion (In the specific sense of having an economic system that requires new resources for expansion but having no new resources to draw upon). In line with our pluralistic approach, this will not be exhaustive of the broader context, but an important part of it, a framing principle.

Below, we briefly outline our framing principle, and then proceed to outline some ways that it might be applied in the case of Mars. The framework is necessarily provisional. It stands in need of fine-tuning. Its application to Mars is also provisional in a further sense: the framework might itself be accepted yet arguments ensue about how best to apply it to any particular planetary surface (or surface of any sort). Up to a point, principle acceptance and principle application are independent. However, the framework for protection as wilderness does seem to have at least one important direct implication (however, it is applied in practice): the default for planetary environmental protection on Mars *ought* to be more robust than the protection of iconic sites such as Olympus Mons or examples of typical terrain, via some system of planetary parks (Cockell and Horneck 2004, 2006). Any plausible system of environmental protection would certainly include these sites. But, if our case holds, the default option could not be restricted to them. There might, of course, be reasons to override the default. Nonetheless, it ought to *be* the default.

10.3 The 1/8th Principle

Our driving consideration is that future generations ought not to be faced with a situation in which there is an expansion dynamic that requires new resources, but the resources of the Solar System have been used up and no new ones can be brought into play. Such a situation might be regarded as a 'Malthusian crisis' in the sense that it would be a case of societally required growth outstripping the physical materials available. Increasingly, marginal resources would have to be brought into play until the point at which no significant new resources were available. The surprise is that such a resource exhaustion crisis may be far closer in time than most of us would expect. This is due to the exponential nature of economic growth and the effective enclosure of the solar system. For the purposes of the argument, we will assume that the Solar System is a closed system. For the most part, this is actually true. Very little comes in from the outside in the form of interstellar asteroids and comets (Trilling et al. 2017).

Given a reasonable ethical concern about future generations, the harms made unavoidable by any such Malthusian crisis are something to avoid. At least, this is the case if the risks of such a crisis are genuine and could not readily be dealt with through some known or predictable mechanism such as a movement of humans out of the system (a possibility set aside at the very start of this paper), or recycling (which would have to exist on a massive scale if we established a stable human presence anywhere else). Of these two, the latter is the most plausible reason for discounting worries about resource exhaustion. However, all recycling systems are imperfect. Their problems of coordination and control will tend to grow with scale, or at least they will grow with the mega-scale of resource utilisation and resource commitment across the Solar System as a whole. For example, if large fractions of the solar system's resources were ever to be built into structures upon which the large-scale continuation of life depended (e.g. O'Neil habitats) this would commit a large amount of resources at any given time. Everything would remain, in principle replaceable, and available to be recycled, but not all at once. Recycling on its own may slow down and mitigate the onset and effects of such a crisis, but if a growth dynamic continues unchecked within a sufficiently large space economy, it will not be able to prevent the arrival of towards a point of 'super-exploitation' (Elvis and Milligan 2019). A point where the new resources required, in order to compensate for the imperfections of recycling, cannot be secured.

The only ways to avoid such a crisis are either to avoid growth on the scale that could threaten it, or else to engage in such growth only on the condition that a safe 'breaking distance', is built-in. That is to say, there must be a point beyond which any deliberate expansion ought to be avoided except in emergencies. We should not intentionally move into the breaking space because it is all that is left before harmful impact. By the time such a point is reached (if it ever is), the economic system should already have transitioned to zero or negative growth. If it has not, then the remaining distance should be *at least* long enough to allow for an emergency transition to occur so that, at the point of new resource exhaustion, any requirement for new resources will have been eliminated or at least minimised (So that the potential for harm is also minimised). In such a scenario, we would still be in trouble, but better placed to deal with it. Our estimate is that even for a modern economic system, it would require at least half a century, i.e. 50 years, for such a zero or negative growth transition to occur. This is comparable to the time from the opening of the first commercial AC electric power station (in Deptford UK) to widespread use (1881–1931), or from the first ARPANET (1969) to today's Internet. Our position, then, is that the limit to deliberate, non-accidental, non-emergency, growth should be *at least* this far away from the final crunch point, the point of super-exploitation.

Given the immensity of the Solar System, it may seem that any concern of this sort is unnecessary. We will never exhaust the system's material wealth. However, we humans have encountered such arguments about limitless bounty before. We have seen the film and do not like the ending. Such claims are similar to those advanced in the nineteenth century to present the Earth as a place of inexhaustible plenty, a great ship sailing through space, amply supplied with many storage holds that might be opened. When one is exhausted, we might simply move to another. This was, as we now appreciate, somewhat naïve. Strictly, it was naïve at the time (George 1912). The failure was not taking into account the *exponential* nature of economic growth. Modest annual expansion leads, cumulatively, to massive expansion. Take, for example, the average growth rate in the West over the period since the beginning of the industrial revolution, a little over two centuries ago. The annual growth of iron usage has been around 3.5%. This too may seem modest and unthreatening. However, an economy growing at 3.5% will double in size every 20 years. After 200 years, 10 doublings, it will be a thousand times larger than at the beginning. While we too are 'at the beginning' in terms of expansion into space, when we allow for the limited available planetary surface, and other available surface, exponential growth at anything like this rate could easily exhaust the system's immense available resources over the course of a limited number of generations. It could do so within the kind of timescales where the actions of the current generation may have a reasonably predictable effect. Beyond a certain threshold, we will allow that there is no way to tell whether or not the overall long-term impact of what we do now will turn out to be advantageous or disadvantageous for those who come later. An epistemic veil falls over the more remote future.

Doubling time goes up quite fast as the rate of growth decreases (e.g. at 2% it changes to 35 years). However, if the intention, hope, or fear, is that a space economy will reach at least a sustained 3.5% growth rate, planetary surface ought to be regarded as a limited resource that could *easily* be exhausted. Given that Mars accounts for such a large proportion of the available surface, caution is required about the Martian surface.

If there is to be at least a 50-year breaking distance, into which we do not want to deliberately move, then zero or negative growth will have to be achieved while two and a half doubling periods remain before the point of super-exploitation. To allow for a little extra room (budgeting for unforeseen difficulties, in the way that any wise project of construction would do) the minimum may well be closer to three doubling periods, totalling 60 years. Adopting this 'tripwire' means that no more than 1/8th of the Solar System's total resources ought to be brought into use. After all, doubling from this point would pass successively through a quarter, half and 100% usage, over only three doubling periods, again at a growth rate of only 3.5%. A higher average growth rate would mean that more doubling periods would need to be added to allow for the half-century emergency slowdown. A lower average growth rate would mean that fewer doubling periods might be required. At present, and given the historical precedent cited, the latter option seems unlikely. However, any requirement to call upon successively more marginal resources might result in a natural slowdown as the economic system grows large and unwieldy. Even on Earth, it is not obvious that the average pace of expansion from the industrial revolution until recent times can be indefinitely sustained.

The assumed figure of around 3.5% growth therefore seems like a reasonable baseline to work from (With allowances for adjustments at a later date). In line with this baseline, we have suggested elsewhere that the adoption of a '1/8 principle' formulated as follows:

'While economic growth remains exponential, we should regard as ours to use no more than 1/8 of the exploitable materials of the Solar System. And by "ours" we mean humanity's as a whole, rather than any particular generation of humans or group of generations. The remaining seven-eighths of the exploitable Solar System should be left as space wilderness' (Elvis and Milligan 2019). Setting aside 7/8 of Mars may seem extreme. It is roughly the inverse of what might be expected in a strict 'planetary park' approach. Yet the justification, given our current economic model, seems strong.

There are, of course, all sorts of workable problems that flow from this position. We might, for example, ask '1/8 of what?' Above, we have focused primarily upon surface area, but the principle applies more generally to mass, volume and resource type. Given that our concern is pragmatic, and that the distribution of resources across the Solar System is extremely uneven (thankfully so, otherwise none of us would be here) it seems best to go sectoral over the issue of measurement and make several allowances in order to respond to the *measurement problem*. Two of the most obvious are allowances for (i) local variations in how resource utilisation is measured and (ii) trade-offs within and between sectors, irrespective of how the lines between them are drawn. The first of these provisions will still allow for a reasonable overall summation of resource use across multiple sectors to occur. There is, in other words, no 'in principle' *measurement problem*, but only a series of smaller problems concerning what the best local measure is going to be in any given case. 15% utilisation by the local unit of measurement in one sector and 5% utilisation, by a different local standard in another sector, will still sum to an overall average of 10%. In the case of planets, such as Mars, surface area looks like a good default for the standard measure, with the interior surface of uncollapsed lava tubes perhaps counting multiply, or else considered not as surface but as a sector in their own right. In the case of Main Belt asteroids, mass or, better still, mass for each asteroid type, may be more appropriate. A summation across the two may still be made.

The second provision, for trade-offs comparable to those for carbon emissions on Earth, will allow for a reasonable flexibility. We acknowledge the imperfections of process associated with terrestrial emission trading; however, imperfections are rarely good grounds for abandonment of a process in the absence of a politically realisable alternative. In the case of Mars, while the default would be 1/8 utilisation, its unique significance could lead us to allow legitimate trade-offs against resources elsewhere (Just so long as any other planetary environmental protection constraints were met). Nonetheless, the default option or starting point should be the reserving of seven-eighths of the planetary surface. This remains so even if we think it likely, or even obvious, that more than 1/8 of the Martian surface is ultimately likely to be brought into use, at some point in time, as a result of political or economic pressures (Or both). Mars has about the same land area as Earth. 1/8 of the Martian surface would be around 18 million square kilometres, This is roughly the same size as South America, a still substantial, continental scale, territory.

10.4 Containment Problems in the Case of Mars

Should the 1/8 principle be applied only at some later point in the process of economic development in space, rather than from the start of mining processes, or at the point when we begin to transform the surface of Mars in non-trivial ways? Assuming, as before, that surface area will be the standard planetary unit of measurement and allowing that Mars accounts for a significant proportion of it, there is a clear case for saying that the clock should already be 'on' when we begin major transformations, if not before. The line between trivial and non-trivial will, of course, be disputed, and there will be borderline cases. But there will also be clear-cut cases: the presence of a lander or rover, on its own, would not count. Strip mining for habitat radiation protection material would do so.

This contrasts with the situation on our home planet, where a variety of social and economic considerations would make any current terrestrial appeal to such a limit difficult to apply. Even the oceans, which cover just over 70% of the Earth's surface, have been significantly modified by our human presence (Jones et al. 2018). Here, on Earth, we are *in medias res*, whereas with Mars we begin with more of a blank slate. And while the utopian dreams of Konstantin Tsiolkovsky for a new and perfect world in space may be a poor guide to what is possible, the admonitions of Ecclesiastes are likely to be just as misleading: there *can* be new things under the Sun. The assumption that we will simply repeat the same old patterns of environmental misuse is not a safe assumption. It separates humans from their relation to history and place. Mars is not another Earth, our human relation to it could not be the same even if we were to strive to make it so. The differences between Mars and the Earth do, however, raise the question of whether or not the planetary surface of the Earth should itself be part of the total surface area of the Solar System that is taken into account for measurement purposes, especially given the large proportion of total available surface area that the Earth takes up. Put in other terms: 'Is the clock already running, now, before any human has gone to Mars?'

Viewing the Earth in the same larger context that we have viewed Mars may incline us to say 'yes'. This is how we have run the position above. However, a 1/8 principle could be run with an Earth excluding approach to measurement, starting with near-Earth asteroid mining or significant human activity on Mars, as soon as it becomes politically viable, i.e. *at the earliest point when it can be applied.* We say this as a reaffirmation of a concession made at the start: there are a range of options for measurement which are consistent with the principle, but are otherwise independent of it. They open up further lines for research and reasonable disagreement.

Given that the clock should, if at all possible, be 'on' when non-trivial transformation of the surface of Mars occurs, how *should* the 1/8 principle actually be applied to the planet when formulated in the terms above? Here, as multiple approaches are possible, we can afford to be a little more speculative than we have been above and supply, on the one hand, clues, hints and pointers and, on the other, a cautionary note about perverse application of the principle. Minimally, on any plausible approach to its application, the principle does point towards the need for a more robust approach than protection via planetary parks *alone* would allow, even if adopted as a way of protecting special or unique sites such as Olympus Mons, or typical areas such as the water-shaped Ma'adim Vallis channel or some of the cliffs with seasonal outflows that may be water and could host life (McEwen et al. 2011). The 1/8 principle is not in any way at odds with the protection of these things, or with the need for planetary parks on Mars. Both would, no doubt, figure in *any* sufficiently robust planetary environmental protection policy. However, protection only of this sort would allow a much larger proportion of Mars to be brought into use without requiring the introduction of any compensating restrictions elsewhere. Quite plausibly the fraction would be reversed, as they are in the United States where 14% of the land area is protected. (A figure close to the 12.5% that is 1/8.)

This is not, of course, what the planetary parks proposal was intended to do. It identified a minimal form of environmental protection rather than an upper limit. Treating it as the latter would be a perverse application of the proposal, deploying it as a way of pressing the claims of 'entitlement to use' rather than of 'duties to protect'.

Safeguards may be required in order to avoid any similarly perverse application of the 1/8 principle, especially in the light of the introduction of a counterpart to carbon emissions trading. As the principle is concerned with a system-level problem of expansion in the face of limited resources, it could be used to override local protection duties, unless suitably constrained. It can do good work, but on its own, it cannot do all of the required work.

This can be seen more clearly if we look at the specific case of applying the principle to Mars. There are likely to be long-term pressures to use more than 1/8 of the planetary surface (perhaps significantly more) and the emergence of such pressures might be beyond anyone's control. Trade-offs would then be required in order to comply with the principle and conform to political realities. But what should such trade-offs look like? It may be tempting to fill in the detail by appeal to some metaphysically deep consideration about the inherent value of some objects and places, and the lack of any such value in other cases, which would then be more suitable for use. The principle *could* be shaped in line with such arguments, and they do have their place. They can help to bring considerations of a deep sort into discussions when appeal to instrumental and pragmatic reasons for protection might miss something important. However, theories of value are unlikely to command the kind of breadth of support that actual policy in a democratic society needs in order to be effective. They figure more in precursor discussions rather than anything akin to policy discussions. And while the analysis here is not exactly of a sort that could be directly adopted as policy (few scholarly arguments are like that), it is at least a little closer to being 'policy apt'.

Accordingly, as made clear from the beginning, we draw upon the human practice of 'valuing', i.e. treating objects and places as consideration worthy in their own right, rather than as having 'value' which would require a deeper metaphysical backstory (The practice is, of course, consistent with such a backstory but does not entail it). Given this, and given that we have already adopted a multigenerational perspective appealing to the interests of future generations, it will make sense to prioritise not only the kind of objects and structures that *we* value, but also those that future generations are likely to value or even, in the case of settlers, to regard as bound into their identity. The Valles and Olympus Mons are obvious candidates. They are striking features of the Martian landscape with potential for a connection to identity at least as much as the Grand Canyon, Uluru, the Great Barrier Reef and the Amazon Rainforest are bound into terrestrial identities.

Less obviously, there are the lava tubes associated with the Martian shield volcanoes, tubes such as those near to Pavonis Mons, or the 'seven sisters', i.e. the caves feeding into the side of Arsia Mons. The mention of lava tubes in this context of a robust protection policy is, however, likely to meet some pushback. They are a key strategic resource on Mars (At least, any tubes which have not collapsed will be of strategic value). They may even be thought of as *the* key strategic Martian resource, the closest resource equivalent to the Peaks of Eternal Light on the Moon (Elvis et al. 2016), albeit for different reasons. While the Peaks are important because of their combination of volatiles and near-constant exposure to the Sun, any Martian lava tubes which have remained intact will be important because of the likely presence of volatiles and, more crucially, their shielding from Solar radiation, which will make them prime settlement sites or prime sites for production operations.

Lava tubes, in a useful condition, uncollapsed or without too much debris and with good protection from intrusions by Martian regolith, may turn out to be so rare and so useful for early phases of human habitation that anything other than near full use would be unthinkable given a sufficiently large human presence. At the very least, they merit inclusion in 'planetary park' protection for a typical sample of what is likely to be lost elsewhere, so that future generations can enjoy the tubes in some approximation to a condition of wilderness. (The claims of science upon any intact tubes would also be considerable with pristine bedrock, regolith-free sites and good conditions for life to flourish.) In terms of a more robust system of planetary environmental protection in line with the 1/8 principle and with an allowance for trading, tubes would almost certainly come under pressure for extensive use. They may be worth counting in a weighted manner as double, triple or greater their interior surface area, rather than something that might be traded-off against comparably extensive areas of the Martian surface.

Protection will thus be more effective in securing the intended outcome (safeguarding valued sites in line with a broader policy of constraint reaching beyond Mars) if other side constraints and/or a system of weighting are involved. But, in order for this to work, and as a disincentive to the repetition of terrestrial mistakes, the trading cost of using lava tubes would have to rise once we move beyond 1/8 and the trading costs of using tubes to the point where pressure is placed upon the final protected examples would have to be prohibitively high. Only considerations of the most serious sort should then lead us to risk their survival in a reasonably pristine condition. The point may be generalised to other special and scarce resources on Mars and elsewhere: trading allowances on their own will offer inadequate protection if we want to protect all that humans are likely to value and identify with. Without weighting and side constraining, trade-offs could be indulged in too easily and the 1/8 principle could be used to give an ethical veneer to rapacious overuse of protection-worthy sites.

Nor are the lava tubes the only sites where special provision would be required in order to avoid perverse applications of the principle. Phobos and Deimos also look like good candidates for weighted protection. As with the larger asteroids such as Ceres and Vesta, they are plausible candidates for treatment as objects with integrity. At the same time, they are also systemically integrated with Mars in a way that might well be salient to the identity of any future human inhabitants. There may even be a case for intervention in order to stabilise their orbits should they erode dangerously or if problems of fragmentation threaten. The same cannot, however, be said of *everywhere*. Inaccessibility, for example, may be an issue. It can shape attitudes. Because of their inaccessibility, particular regions may never feed quite so directly into Martian identity in a strong way. However, inaccessibility is not everything. The Martian poles may be a partial counterexample to reliance upon accessibility alone to make an assessment of protection worthiness. They may be difficult to access (perhaps not as difficult as once assumed) but their iconic status could well feed plausible calls for protection. There are obvious parallels with the

Earth here. Protection of our terrestrial polar ice caps is not exclusively driven by concerns about global warming. The polar caps have come to have a significance in their own right.

In the longer run, Mars need not be seen as a closed system. Martian resources may also be traded-off against resources elsewhere, e.g. the Main Belt, with suitable weighting. And, the greater the trade-offs between utilisation of Martian resources and of resources elsewhere (e.g. in the Main Belt), the stronger the case may be for using access to both as a stepping stone to the Kuiper Belt, access to which could then legitimate a greater use of a ten times larger amount of resources than we might access in the inner area of the Solar System (Gladman et al. 2001), without automatically violating the 1/8 principle. The bigger the pool of accessible resources, the bigger the 1/8 will be.

However, even with an assumption that we *may* ultimately access the Kuiper Belt, caution is required to avoid too many trade-offs between *distinctive* sites in the inner Solar System, near to Earth, and larger amounts of more *mundane* asteroid materials elsewhere, in either of the belts. Such an application of the principle truly would be perverse, in the sense of legitimating an indefinite postponement of environmental protection and licensing short-term destruction of the most obviously protection-worthy locations, just so long as promises of large-scale future constraint are made.

10.5 Conclusions

It is better that good principles are in place, rather than principles of any other sort. However, good principles may be applied badly and there is ultimately no safeguard against the latter beyond the safeguards offered by culture (political, ethical, social). There is no clever way to render any principle 'self-protecting' against misuse. No such guarantees can be offered. What we can, however, do is draw attention to two key considerations. First, there is a need for some overall growth principle in response to the fact that exponential growth goes from minor use to full exploitation rapidly. Second, the proper application of such a principle should bear in mind its driving rationale *as* a protection measure, as a safeguard against runaway economic development, and not a legitimation of such development.

In its application to Mars, we have highlighted a number of key points: the planet is a unique place in the solar system; its resources are finite, though large, and constitute a significant proportion of the available resources of the Solar System; in the case of some strategic and particularly limited resource, competing claims may be anticipated; protection of these and other Mars resources ought to go beyond traditional 'planetary protection' to include 'planetary environmental protection'. The 1/8 principle, although of broader scope than concern with Mars alone, seems able to accommodate considerations of this sort. Or, more formally, it allows us to do justice to the particularity and uniqueness of the planet. This counts in its favour. Any overall growth principle which yielded radically counterintuitive proposals when applied to Mars or to any other candidate site for human use, including habitation, would be problematic. The 1/8 principle does pose significant political, ethical, and social challenges but is nonetheless a prudent boundary and organising idea. It is relatively easy to turn into policy options, and can be applied to subcategories and aggregated using weighting schemes in line with human practices of valuing and use. These considerations may not be decisive, but they do count strongly in its favour. It has the key features that any plausible overall growth principle must have.

References

Cockell, C. S., & Horneck, G. (2004). A planetary park system for Mars. *Space Policy, 20,* 291–295.

Cockell, C. S., & Horneck, G. (2006). Planetary parks—formulating a wilderness policy for planetary bodies. *Space Policy, 22,* 256–261.

Conway, E. M. (2015). *Exploration and engineering: The jet propulsion laboratory and the quest for Mars*. Baltimore: Johns Hopkins University Press.

Elvis, M. (2014). How many ore-bearing asteroids? *Planetary and Space Science, 91,* 20–26.

Elvis, M., & Milligan T. (2019). How much of the solar system should we leave as wilderness? *Acta Astronautica.*

Elvis, M., Milligan, T., & Krolikowski, A. (2016). The peaks of eternal light: A near-term property issue on the Moon. *Space Policy, 38,* 30–38.

George, H. (1912). *Progress and poverty: An inquiry into the cause of industrial depressions and of increase of want with increase of wealth.* New York: Doubleday.

Gladman, B., et al. (2001). The structure of the kuiper belt: Size distribution and radial extent. *Astronomical Journal, 122,* 1051–1066.

Green, J. (2017). Making Mars livable. https://www.youtube.com/watch?v=58RPasKQlwA. Accessed October 8, 2018.

Jones, K. R., Klein, C. J., Halpern, B. S., Venter, O., Grantham, H., Kuempel, C. D., et al. (2018). The location and protection status of Earth's diminishing marine wilderness. *Current Biology, 28*(15), 2506–2512.

Lupisella, M. (2016). Cosmological theories of value. In J. Schwartz & T. Milligan (Eds.), *The ethics of space exploration* (pp. 75–91). Heidelberg, London and New York: Springer.

McEwen, A. S., Ojha, L., Dundas, C. M., Mattson, S. S., Byrne, S., Wray, J. J., et al. (2011). Seasonal flows on warm martian slopes. *Science, 333,* 740–743.

McKay, C. P. (2013). Astrobiology and society: The long view. In C. Impey, A. H. Spitz, & W. Stoeger (Eds.), *Encountering life in the universe: Ethical foundations and societal implications of astrobiology*. Tucson: University of Arizona Press.

Milligan, T. (2015). Asteroid mining, integrity and containment. In J. Galliot (Ed.), *Commercial space exploration: Ethics, policy and governance*. Famham: Ashgate.

Milligan, T. (2018). Valuing humans and valuing places: "Integrity" and the preferred terminology for geoethics. *Geosciences, 8*(1), 25.

O'Neill, G. (2001). *The high frontier: Human colonies in space*. Ontario: Apogee.

Rawls, J. (1974). The independence of moral theory. *Proceedings and Addresses of the American Philosophical Association, 47,* 5–22.

Schwartz, J. S. (2018). Myth-free space advocacy part III—The myth of educational inspiration. *Space Policy, 43,* 24–32.

Taylor, A., Elvis, M., & McDowell, J.C. (2018). *Acta Astronautica*, 146, 73–82.

Trilling, D. E., Robinson, T., Roegge, A., Chandler, C. O., Smith, N., Loeffler, M., et al. (2017). Implications for planetary system formation from interstellar object 1I/2017 U1 ('Oumuamua'). *Astrophysical Journal, 850,* L38.

Tsiolkovsky, K. E. (2004). *Selected works of Konstantin E. Tsoilkovsky*. Honolulu: University Press of the Pacific.
Turner, F. J. (2015). *The frontier in American history*. Michigan: University of Michigan.
Zubrin, R. (2012). *The case for Mars: The plan to settle the red planet and why we must*. New York: Simon & Schuster.

Chapter 11
Science and Ethics in the Exploration of Mars

Gonzalo Munévar

Abstract The scientific exploration of Mars might yield results of extraordinary importance for our own planet, particularly the search for extant or fossil Martian life, which would make it possible to understand terrestrial life in a more profound way. This potential scientific treasure places on us an ethical obligation to minimize the disruption of the Martian environment until our scientific exploration has been greatly advanced. We also have ethical obligations to the human scientific explorers of Mars, ethical obligations that require a series of scientific investigations, e.g., about how the low Martian gravitation may affect those explorers' physiology.

11.1 Introduction

This chapter will discuss the interplay of scientific and ethical considerations relevant to the exploration of Mars. I will be concerned both with such interplay in terms of our relationship with the red planet and in terms of the well-being of the scientists doing the exploring.

There are many excellent reasons for the scientific exploration of Mars. Learning about its geological and atmospheric history could give us important insights about the history of our own planet, which in turn can give us clues about the Earth's future. But the most important potential discovery concerns Martian life, whether extant or fossil. Having another form of life to compare with terrestrial life would be of extraordinary value. Until this exploration is pretty much complete, it would be rather irresponsible, indeed unethical, to change the Martian environment in any significant way, which a large colonization is likely to do.

Since the scientific exploration of Mars is likely to demand long tours of scientific duty, given the enormous costs and distances involved in placing and equipping human scientists on Mars, intensive research efforts must be undertaken as well to minimize the potential harm to humans in the long stays involved.

G. Munévar (✉)
Lawrence Technological University, Southfield, MI, USA
e-mail: gmunevar@ltu.edu

K. Szocik (ed.), *The Human Factor in a Mission to Mars*,
Space and Society, https://doi.org/10.1007/978-3-030-02059-0_11

That research, as well as the knowledge to come from our scientific study of the planet, will make future colonization efforts more likely to succeed. At that point, many practical reasons for colonizing Mars will become apparent. One is to give humanity and the terrestrial life we take with us to Mars a chance to avoid extinction in case that a cosmic catastrophe destroys much of the Earth's biosphere (e.g., collision with an asteroid much larger than 10 km across). A second is the many inventions that would come from trying to develop large human settlements on another world, just as the space program itself produced many inventions and stimulated the use of many others. A third is the experience itself of adapting to another world, for which our descendants in the long future will thank us (Munevar 2014).

11.2 The Value of the Scientific Exploration of Mars

Hypotheses about the Earth are inevitably tied to more general theories about the nature and behavior of the other bodies of the solar system. As we challenge our understanding of that system, we place ourselves in a position to learn new things, not only about other worlds, but also about our own. The bounty of space science will thus not be scattered by alien winds over alien lands. It will be handed down to the children of the Earth. This is one reason why a scientific opportunity to explore Mars imposes on us an additional ethical obligation not to squander it.

As an example of how trying to understand our planet benefits from comparative planetology, consider that the Earth's gravitational energy keeps its atmosphere from dissipating into space, and thus it determines to a high degree the density of that atmosphere. That density in turn influences the chemistry of the environment and the climate of the planet. To see the point clearly it pays to compare our planet with others. For example, the density of Mars' atmosphere is so low (about one hundredth that of Earth's) that water cannot exist in liquid form: It goes directly from ice to vapor.

The Moon has influenced terrestrial climate in at least an important way: Its gravitational influence stabilizes the tilt in the Earth's axis of rotation so that it varies only a few degrees. Mars, by contrast, may have suffered wild swings in the tilt of its axis, and this instability might have had devastating consequences for the Martian climate (Toon et al. 1980). In other words, the Moon may have played a crucial role in ensuring that life on Earth endured and prospered while Mars became a barren world.

In investigating other worlds we find:

(1) Valuable information that serves to refine our theories of the origin and evolution of the solar system, and hence of the Earth.
(2) Unusual phenomena that stretch our views of basic terrestrial mechanisms.
(3) Opportunities to test our ideas about the Earth—the solar system serves as a natural laboratory.

In the standard account of the evolution of a rocky planet, the denser the planet, the more heat it will have available from radioactive elements; and the larger the planet, the more retarded the loss of heat. By this account, no planets much less massive than the Earth could still have active volcanoes or relatively young surfaces. Mars, for example, shows evidence of recent volcanism (within the last two million years); but even if Mars is not a dead planet, its surface is testimony to a prolonged coma.

It seems, however, that earlier in its history, when Mars' internal energy was much higher, Martian volcanoes might have filled the atmosphere with perhaps 100 times as much CO_2 as today (Forget et al. 2013). This factor would have raised the density and temperature enough to permit liquid water and large amounts of water vapor, and hence much more of a greenhouse effect. As we look at Mars now, that appears to have been the case. For years, spacecraft photographs showed what seemed to be riverbeds and suggested other indications of significant amounts of liquid water in the past, perhaps even an ocean. More recent evidence, discussed below, indicates at least one large deposit of underground liquid water.

This view of Mars is strengthened by the recent discovery that Mars at one time did have at least the beginning of plate tectonics as well, and may experience "Marsquakes" about every million years (Yin 2012). It seems, then, that life could have existed on Mars, and may still. If so, why did not Martian life control the climate the way Earth's presumably did? Mars apparently did not have enough energy to run the cycles that have made a sustainable biosphere on Earth possible, just as it did not have enough heat to support the motion of tectonic plates for billions of years. Life, if it ever existed on Mars, was thus powerless to stop the ultimate collapse of its global environment. Earth was fortunate not to be besieged by such extreme conditions.

What would Martian life enable us to learn about our own kind of life? The answer is clear. All life on this planet is based on the same carbon chemistry and apparently, all have the same genetic code. Of the many possible amino acids, only 20 are used to build proteins. DNA, the reproductive code for terrestrial life, makes use of only four bases. Moreover, organic molecules can be left handed or right handed, but terrestrial life prefers left-handed amino acids and right-handed sugars. Are these circumstances mere accidents of organic evolution, or are there fundamental reasons why life has taken these particular turns on this planet? Even *one* other kind of life would permit us to make great strides in examining these matters. For that, other life may use a wider range of amino acids and bases, or it may prefer right-handed amino acids or left-handed sugars. One result of such a finding may be that, say, a particular chemical balance in the Earth's oceans caused the preference for left-handed amino acids. Or the alien life may be similar to life on Earth, which would reveal to us some sort of organic inevitability. The new perspective would be very fruitful in trying to understand our own biology at all levels.

It would be especially useful to observe stages of organic evolution and to study life as it begins in a new world, or at least to find fossil records of such beginnings. Even if perchance organic evolution produced in two similar planets similar primitive cells with essentially the same genetic code, the subsequent evolution would have much to teach us, for life in those two planets would undergo different histories of adaptation. Imagine, for example, that the now famous Alvarez asteroid had not crashed on the Earth. No one knows how dinosaurs would have continued to evolve, but it is possible that their grip on the surface of the planet would have been further strengthened. Mammals might have been thus forever condemned to crawl and scratch in the night like so many other vermin.

Even similar planets are likely to exhibit different tectonic histories. Plate tectonics brings continents together or breaks them apart; it throws chains of mountains up over the landscape; and it creates volcanoes where the plates rub against each other. In doing so, it brings some habitats to an end and others into existence. It destroys. It influences. It changes life in many ways. Slight differences at the beginning of the history of a planet would alter the make-up of the crossroads that life has to face, first at the level of organic chemistry, and then at the level of cells—presuming that cells are common to living things. A eukaryotic cell (a cell with a nucleus) may well be the result of symbiosis between different varieties of prokaryotic cells (without a nucleus) (Bylinsky 1981). For example, the mitochondria in eukaryotic cells may be the remnants of prokaryotic cells that discovered how to use oxygen for energy and were swallowed but not digested by larger bacteria. Since eukaryotic cells are the building blocks of all complex organisms on Earth, we can imagine that different symbiotic relationships between primitive cells might have led to forms of life vastly different from those of our acquaintance. On planets so endowed, the subsequent interaction of life with the rest of the environment would have a multiplier effect, for they would change their environment in novel ways, and those new environments would lead life to adaptations that on Earth could meet only with misfortune.

Acquaintance with such alternative biotas would inevitably lead to profound transformations in biology, since biology would grow, and scientific knowledge seldom grows without changes. In this, scientific knowledge resembles animals. Mammals, for example, did not just get bigger after the extinction of the dinosaurs. As their size increased, the structure of their skeletons had to change to accommodate their larger weight. In a planet with gravity similar to ours, a dog the size of an elephant would probably look much like an elephant. In an analogous manner, a science of biology that were suddenly much larger in subject matter would have to grow connections and supporting structures for which there was little need in the days of a single biota.

On Mars, the unprotected surface is not the most promising place to look for life. There are sites in Antarctica devoid of life on the surface, but if we care to dig we may find it in porous rocks below. Some raise the possibility that Mars might offer extreme forms of life. There are numerous examples of life surviving under extreme conditions: in the core of nuclear reactors, in underground streams with temperatures of hundreds of degrees Fahrenheit, under incredible pressures at the bottom of the ocean. Organisms have been found even deep in the Earth's crust. No extreme habitat, though, is as challenging as the Don Juan Pond in Antarctica, where the salinity is so great that a random sample is likely to fail the Viking test for the presence of carbon compounds (NASA's two Viking landers used tests that, in combination, failed to find life in Mars). But living organisms exist in the Don Juan Pond! (Siegel 1970,

1968; Siegel and Spettel 1977; Siegel and Siegel 1980). And they can also survive in space (Jönsson et al. 2008).

Of course, the Viking experiments were designed to detect *average* life on the surface, or inches from the surface, whereas it is clear that if any life exists on Mars it should be in *extreme* conditions, or at least in protected habitats probably not very close to the surface.

If Mars fails us, life might have nevertheless made a stand in the organic clouds of Jupiter or perhaps in some underground caves in active Io. Another moon of Jupiter, Europa, is covered by smooth ice, which indicates a good amount of internal heat, and probably an ocean of water under the ice. Another notable prospect is Titan, the large moon of Saturn with a dense atmosphere and at least traces of organic compounds. Unfortunately, Titan is too cold for life as we know it—cold enough (−288 F°) that the argument about the ability of life to adapt to extreme habitats begins to wear thin.

At any rate, the "extremophile" evidence can show only that once life begins it can adapt to very hostile conditions. But it does not show that life could begin in such conditions. These are two very different things. Let me illustrate this point by means of an analogy. During pregnancy, many substances can be lethal to the developing embryo, e.g., alcohol, tobacco, and hallucinogenic drugs. The chances for the new life are greatly hampered under those conditions. Once the baby is born, the situation begins to change. Eventually, it may grow into an adult who smokes, drinks and abuses drugs, none of which are conducive to a healthy life, but none of which need be immediately lethal either, as they could be to the embryo. Furthermore, life might not be able to make a start on a planet that would otherwise be exactly like today's Earth, but it surely has no trouble flourishing in it now.

As we saw earlier, in a thin atmosphere, such as the present Martian atmosphere, water goes from solid to vapor without first becoming liquid. Therefore, the surface features that resemble river deltas and suggest past running water in turn constitute evidence that the atmosphere was much denser once upon a time.

There are alternative hypotheses on the dendritic channels, though. According to one of them, for instance, occasional but pronounced tilts in Mars' axis of rotation would expose one of the poles to the full action of the sun. If that were the case, the melting polar cap would provide enough pressure to permit water to run off and presumably form those surface features—all without the benefit of a dense atmosphere. Photographs by the Mars Reconnaissance Orbiter, with ten times better resolution than any taken before, indicate that lava and wind-driven dust have run through those presumed river channels and gullies far more recently than water, even though catastrophic floods might have carved them once upon a time (Science 2007).

Nevertheless, the preponderance of new evidence indicates that the Martian atmosphere was far denser once upon a time and that liquid water ran on the Martian surface. The most striking findings are those of Opportunity, a Mars Exploration Rover. Opportunity found, for example, salt deposits in a region called Meridiani Planum. According to Mike Carr, the man who wrote the book on water on Mars, it is clear that a large body of water existed in that region. It is also clear that the "water had to pass through the ground to pick up the dissolved ions that ultimately were precipitated out as salts" (Carr 2004). This means that the water in that region (a lake or a sea) could not have been mere runoff from melted polar ice. At any rate, the positive side of the case for water was greatly strengthened by the discovery of a 20 km wide lake of water about 2 km under the ice of the South Pole. The discovery was made using radar measurements from the Mars Express spacecraft (Orosei et al. 2018).

If we assume that planets similar to the Earth have similar beginnings—in this case, similar distributions of organic materials, atmospheric gases, and sources of energy—and if we keep in mind that our own earliest fossils are about 3.5 billion years old, it seems plausible to suppose that life made a start in Mars. If that is so, the possibility exists that in some regions of Mars, we may find fossils of organisms that thrived in days when the atmosphere was denser and warmer.

Although such Martian fossils would perhaps not be as exciting as living organisms, they still would be invaluable in that they would permit us to compare our form of life with an alien one. We may also be able to draw some interesting lessons from a failed interaction between life and a planetary environment.

As luck would have it, though, liquid water is likely to exist in underground deposits. The permafrost is presumed to exist to a depth of hundreds of meters, which suggests that some liquid water may be found in proximity to magma deposits and other sources of thermal energy. Martian life forms, if any exist, need not be quite as extreme after all.

11.3 The Value of Astrobiology with or Without Specimens

The lack of extraterrestrial specimens is an objection to the pursuit of astrobiology only if we accept a narrow definition of the field. Astrobiology goes beyond the search for extraterrestrial life: It is largely the application of space science and technology to understand how life may originate and evolve anywhere in the cosmos. As a practical matter, astrobiology often devotes itself to investigating how life originated and evolved on *this* planet. Astrobiology tries to determine, for example, what the Earth was like 3.5–4.5 billion years ago—what was the ultraviolet flux? What were the volcanic and other tectonic activity? How much molecular oxygen was in the atmosphere? And how much ozone? How much carbon, hydrogen, and nitrogen were "recycled" through the Earth's crust and how much were brought to the Earth by asteroids and comets? To decide these issues, we must go away from the Earth to study the older surfaces of the Moon and Mars, the presumably still primordial atmosphere of Titan, and the largely untouched chemistry of comets.[1] Astrobiology is thus inseparable from comparative planetology.

This connection is all the more evident when we remember that to understand the nature of an environment well we need to understand its origin and evolution. In

[1] Although the emphasis has been mine all along, in making these remarks I find myself paraphrasing Harold P. Kline's many comments on earlier drafts of this essay.

the global environment of the Earth, life has played a crucial part, by changing the chemical composition of the atmosphere, its density, and its temperature. How life originated is thus a question of great importance if we are to understand how our global environment came to be as it is. At the same time, we cannot begin to settle that question without making some critical determinations about how the planet was formed, how its atmosphere was created, how much energy it received from the sun in its early evolution, and in general all those questions that form integral part of comparative planetology.

In trying to answer the question of the origin of life, however, there are great difficulties of substance and of method. For example, some investigators require explanations that make life somehow inevitable or at least very likely. Given the early conditions in the planet (e.g., a reducing atmosphere and later a primordial soup of organic materials), and processes that should be expected (e.g., radiation, lightening), organic evolution towards life should be highly probable. Another school of thought would have a series of extraordinary coincidences bring life about. Thus, even if organic matter was abundant on the early Earth, it would have taken an accident, or accidents, to get organic evolution on the road to life. To illustrate the sort of disputes involved, let me consider the hypothesis that a large moon is needed for life to begin. Of course, if that hypothesis is right, we should conclude that life is probably very rare in the galaxy and not to be found in the solar system at all, except for our own kind.

According to this hypothesis, clays in shallow waters served as the templates for amino acids to combine into the first complex organic molecules by forming peptide bonds—bonds that can link carbon and nitrogen in separate organic chains. Such a bond can form when, for example, a nitrogen atom loses its bond with a hydrogen atom (H) and a carbon nearby loses its hydroxide bond (OH). Since the H and the OH combine to form water (H_2O), the formation of the peptide bond requires a loss of water. The problem is how to lose water in the presence of all that water in which the amino acids are suspended.

The tides created by the Moon provide the solution to the problem (Dorminey 2009). When the tides go out a residue of amino acids is left on the clays, and the heavier concentration in drier surroundings allows the peptide bond to form. Once formed, the peptide bond is stable in water. After many repetitions of this process, organic molecules of an increasing complexity can be formed. The function of the clays is to provide a mechanism for replication.

Having a day–night cycle also seems important because the opportunity to move away from equilibrium gives the prebiological molecules a chance to vary, and this opportunity for variation is an essential characteristic of evolution. Cyclical events are in general favorable because they permit the molecules to reach a state of equilibrium to consolidate their gains before having to change again. Apart from the tides and the day–night cycle, we have the concomitant temperature fluctuations. And we have seasons because the Earth's axis is tilted in just the right way thanks, again, to the Moon. Then we must also take into account the role of the magnetosphere and many other factors whose possible relevance or even their very existence may escape the experts at this time.

Insofar as any of these are large factors in allowing life to gain a foothold on Earth, the origin of life becomes an improbable event. But we simply do not know. As plausible as hypotheses such as this may seem today, they may sound very quaint in two or three decades, let alone in a few centuries. And even if the events in question were indeed factors in bringing life into our world, on further examination, they may turn out to be just some among the many alternative mechanisms that could have provided for the evolution of evermore complex molecules in a variety of other worlds. It happens all too often that when a mechanism cannot be immediately proposed to explain a particular step on the way to life, people who ought to know better jump to the conclusion that life on Earth was an extraordinary coincidence.

Many biochemists, for example, have felt that the problem of the origin of macromolecules is insoluble. At the most basic level, life consists of nucleic acids (such as DNA and RNA) that contain the genetic information, and of functional proteins (such as enzymes). If we imagine that DNA or RNA was the original macromolecule we have to explain how it could replicate in the absence of enzymes, which are essential in modern living systems. On the other hand, if we imagine that the proteins came first, how could they have built around themselves the nucleic acids that would carry the information necessary for future coding of the same proteins?

For years, none of the mechanisms proposed seemed satisfactory, not even co-evolutionary mechanisms, because the chemical association of nucleotides (the building blocks of nucleic acids) and amino acids (the building blocks of proteins) was just too problematic. On the face of this situation, some people thought that life was a stroke of luck. And some others even suggested that life had probably come from elsewhere to the Earth (the so-called "Panspermia Hypothesis"), as if removing the problem of the origin of life a few light years amounted to a solution (de Duve 2002, 2005).

Nevertheless, some plausible mechanisms were proposed in the late 1960s and have been refined ever since. One of them, suggested by A.G. Cairns-Smith, is that microscopic crystals in clays can serve to replicate molecules. Such clays have a large capacity for adsorption, which causes tiny bits of proteins to stick to them, just as particles of meat do to the surface of a frying pan. The crystals in these clays would then grow and reproduce the patterns of the amino acids adsorbed in the clays (Cairns-Smith 2009).

These processes can be repeated millions of times, until with the development of enzymes, as J.D. Bernal notes, we would also see the appearance of co-enzymes, some of which are identical to the nucleotides of RNA. As the co-enzymes are adsorbed, their efficiency in chemical energy transfer would give clear reproductive advantages to their associated enzymes. Under these conditions, a co-evolution of functional proteins and nucleic acids becomes possible. And this result presumably paves the road to the eventual origin of the first cells. This view has been buttressed by the work of the space scientist James Lawless, who has shown that clays do select precisely the amino acids that can form biologically active proteins.

There are many other hypotheses buttressed by experimental work that fill in some of the steps deemed necessary to take us from atmospheric gases to living cells (e.g., proteins that can "make" their own RNA). But, what is necessary and what is not depends on the approach one takes to explain the origin of life. First, there are different starting points. Some want, even demand, a reducing atmosphere (poor in oxygen, rich in hydrogen and other gases like methane). Others think that the original atmosphere was composed largely of carbon dioxide. Second, then comes a story of the evolution of organic matter, a story that may involve thousands of steps, of which only some are specified. And of course, there could be alternative plausible stories. What makes them plausible is that some of the steps that may have seemed baffling at one time can be produced in the laboratory now, while others can be explained theoretically. For example, the first generally accepted story gained its plausibility from the Miller-Urey experiment, in which a reducing atmosphere in a flask was subjected to electrical discharges. Presumably, the result of the experiment was a soup containing the building blocks of life. Apparently, however, only very few interesting organic molecules were actually produced in such an experiment, and those were of very little complexity. The steps from there to, say, a self-replicating molecule, let alone a cell, are truly gigantic.

Since there are so many ways to tell the story, most equally unconstrained by the scant evidence, it is not surprising that the intuitions of different investigators differ on what is crucial and what is not. And even if most of the apparently necessary steps of a particular story can be accounted for by experiments, there remains the difficulty that the answer to one part of the puzzle is often at odds with the proposed answer to the next part, e.g., a molecule used as a building block for a more complex molecule is produced in an alkaline solution, but the more complex molecule has to be produced in its opposite, an acidic environment; this is not a fatal setback, since in living things the product of a reaction can be transported to a different internal environment to be used to build something else, and in general we find that natural processes have co-evolved in the living world to accomplish just this transport. The problem is that it all seems just too convenient. What we want to know now is not just how it could have been, but how it was—we want the "real" story.

To go from just-so stories to compelling hypotheses, we need a better understanding of the initial conditions on the Earth, and as we develop our hypotheses accordingly, we will get ideas of what sorts of evidence about the subsequent evolution of the global environment we may want to look for. One helpful way to proceed is to examine those worlds where life might have started, or at least where we should expect some small amount of organic evolution. To the extent that organic evolution has taken place there, we learn much about our own, once we factor in the relevant differences. To the extent to which organic evolution has not taken place, we also learn much about the failure of some forms of reasoning about the origin of life and perhaps get some clues about more appropriate forms of reasoning.

Astrobiology and comparative planetology will merge in many other contexts. Take, for example, the search for the origin of the organic carbon on the Earth, surely a needed background to make a definitive determination of how life started on the Earth. To have organic compounds, we first need to trace the carbon and the other relevant elements (hydrogen is normally easy, since it is almost everywhere, with exceptions such as the Moon and Mercury). We begin the search for carbon in the solar system and then see how it was apportioned to the Earth. If the Earth had a disproportionate amount of carbon, we must deal with a certain set of scenarios in which the Earth comes to occupy a privileged position. If carbon is very common in the solar system, as indeed it is, our scenarios are of a different sort, but we still want to know how the Earth came by the amounts that it has: Did it happen during the initial accretion of the planet, or was most of the carbon brought in by the subsequent bombardment by comets and asteroids? We might get some useful clues by examining Mar's history.

These and other investigations underscore the intimate connections between astrobiology and those aspects of space science that deal with the formation and evolution of planets. Since the role of life has been of crucial importance for the Earth, and since we need to know specifically how terrestrial life may not only survive but also prosper, this study of origins is highly justified. To put the point differently, the biota and many of the other elements of the global environment co-evolved. Thus, to understand the evolution of one of those elements we need to understand it in its relationship to the evolution of the others. Moreover, the very role of the imagination in trying to determine the range within which life can be born and the possible forms life may take provides a fruitful context in which to discuss questions of origin and evolution. For by the consideration of likely scenarios for life, and by the comparative examination of the planets in our solar system and of other planetary systems, we will be better able to understand not only how life came about but also why it took the paths that it did when it apparently had others available.

11.4 The Value of Searching for Martian Life Even if no Specimens Are Found

As we have seen, even if we find no life in Mars, the search may still help determine how life can be born and the possible forms it may take. But Even *presumed* dead ends, such as the Martian meteorite ALH84001, have spurred fruitful biological investigations. Martian Meteorite found in Antarctica in 1984 had arrived 13,000 years earlier, after being blasted from Mars some 17 million years before. It had crystalized perhaps about 3–4 billion years ago, when there was liquid water on Mars.

In 1996, David McKay, of NASA, claimed evidence of fossil life in ALH84001 (worm-like structures in the center of the meteorite, 20–100 nm long, although some were larger). In close proximity, he also found globules of carbonate, polycycle aromatic hydrocarbons (PAHs), clear evidence that there has been organic carbon in Mars, as well as magnetite and iron sulfides (McKay et al. 1996). Meteorite experts and many other scientists disagreed with McKay's conclusion (Kerr 1997)[2] on the grounds that:

[2]The controversy spread to public arguments in the newspapers; see for example the front-page article "Life on Mars: Scientists 'thrilled' by prospect," *Seattle Times*, August 7, 1996.

(1) All the compounds and structures found in ALH84001 could have been produced by inorganic processes.
(2) Therefore, by Occam's Razor, we should eliminate the extraordinary conclusion (Martian Life).
(3) Moreover, the smallest worm-like structures were up to 50 times smaller than Earth bacteria (not enough space for all the materials a cell needs).

A number of replies were made to the majority arguments:

- To (1): Inorganic magnetite forms at three times the temperature that ALH84001 experienced. And the magnetite in the sample is extremely pure (on Earth only bacteria make it that pure).
- To (2): In ALH84001 McKay found three things in space a few nanometers across: (a) typical bacterial food (hydrocarbons), (b) structures that look like typical bacteria, and (c) typical excreta of bacteria (magnetite and iron sulfides). It thus seems to me that one simple hypothesis, **LIFE**, accounts for all these phenomena and the fact that they are closely packed together: Martian bugs ate the hydrocarbons and left the droppings behind. On the other hand, the inorganic-origins hypothesis requires at least three separate mechanisms and has little to say about why they are together in such a small space.

At any rate, the controversy spurred interest in the possible existence of undiscovered Earth bacteria that small (a subject of very little interest in the 80s and early 90s). Soon enough some biologists claimed to have found many such varieties of bacteria (nanobacteria), even smaller than the presumed Martian fossil bugs. Later investigations revealed, however, that main nanobacteria candidates are actually nonliving mineral structures, e.g., calcium carbonate crystals, which mimic bacteria in some respects and even reproduce (Martel and Young 2008). This was a very interesting discovery with practical importance: Those nanostructures are apparently involved in the formation of kidney stones.

The search for ultrasmall bacteria did not stop there. In 2015, scientists from the University of California at Berkeley and the Lawrence Berkeley National Laboratory announced in *Nature* their discovery of ultrasmall bacteria, about the size of the larger Martian worm-like structures (250 nm). And just to be sure, they also *sequenced* the ultrasmall bacteria genomes (Luef et al. 2015). These ultrasmall bacteria are four times smaller than life "could be" (according to the previous common "wisdom"). Thus, the pursuit of the possibilities suggested by a "failed" hypothesis led to significant biological discoveries. Consider now that the primitive Martian RNA bacteria (without DNA) could be even smaller: About two or three times smaller than ultrasmall bacteria. That is, they could be the size of the worm-like structures in ALH84001! On the extremely dynamic Earth, such fossils would be far older than those of the DNA bacteria we have found (less than four billion years old). But on a far less dynamic Mars they might have been preserved. Finding them would be far more than exciting.

The vigorous search for life in Mars would inevitably lead to the profound transformation of our views of the living world. And since those views are linked to our understanding of the global environment, the resulting theoretical adjustment would be of great magnitude—and so eventually would be the change in the way we may interact with the universe. Whether or not we ever find a single Martian specimen!

11.5 Science and the Welfare of Human Explorers in Mars

Steve Squyres, a robot expert and the lead scientist on the Martian rovers, said in 2009 that

> … the most successful exploration is going to be carried out by humans, not by robots. What Spirit and Opportunity have done in 5 1/2 years on Mars, you and I could have done in a good week. Humans have a way to deal with surprises, to improvise, to change their plans on the spot (Squyres 2009).

Human scientific explorers in Mars will arrive there after a long and expensive trip and will require a very sophisticated infrastructure to survive, let alone thrive in their work. It is reasonable to suppose, then, that their tours of duty should be rather long, years perhaps. One concern that arises from any human presence in Mars is the possibility of contamination, both of Mars by terrestrial life, particularly microbial, and of the human outposts by extant Martian life. The first is, of course, a greater concern, though the planetary surface is so inhospitable as to make it unlikely that even extremophile may just barely survive in pockets. Interested readers may consult the latest in-depth discussion of planetary contamination in *Review and Assessment of Planetary Protection Policy Development Processes* (National Academies 2018).

The other concern that arises from the human exploration of Mars is the protection and well-being of those human explorers. The thin Martian atmosphere does not offer much protection from radiation, a situation made worse by the lack of a significant magnetosphere. Living spaces, vehicles, and special outdoor suits, however, can be built to offer the needed protection. Air, fresh food, and comfortable temperatures will also be available. The low gravity of the small planet, about 1/3 that of the Earth, may become an obstacle to very long stays. Perhaps the problems will be far less significant than they are in microgravity, e.g., in the Space Station, but it would be highly irresponsible not to investigate the potential ill effects. In the rest of this essay, I will describe some of the relevant studies of the biological effects of microgravity and make some remarks about possible extrapolations to effects on humans on Mars.

Our main interest lies in the range between 0 and 1 g, so as to study not only the perception of gravity but perhaps even the role that gravity has played in evolution. By experimenting in that range, we may be able to determine gravitational thresholds of biological importance; that is, we may determine the minimum level at which gravity can be detected and at which it becomes a significant factor in physiological or developmental functions. With some luck, the gravitation humans will experience on Mars will be above that threshold.

Gravity is all-pervasive in our planet; it is not hard to imagine that life took advantage of its presence to favor some avenues of evolution over others. As Galileo noticed as early as 1638, how much an animal weighs depends on how well its bones can support it.[3] Thus its anatomical structure depends on gravity. In a planet with lower gravity, we may find much taller animals and more symmetrical trees (the symmetry is often broken because slight differences in mass in the branches weigh the tree down in different ways; the more the gravity the more pronounced those differences become in the development of the tree).

As for the perception of gravity, it greatly influences the way a plant grows. In the microgravity of space, roots grow out of the ground into the air and the shoots are generally disoriented in spite of the constant illumination from the top. Gravity is obviously the main factor in the case of the roots; shoots require both gravity and illumination. But how do plants recognize gravity? They have gravity receptors, and thanks to space research we are beginning to understand what those receptors are.[4]

So far there are indications that gravity may also play a role in the axial orientation of amphibian embryos (which is a factor in the normal development of amphibians) and perhaps also in that of birds (Kochav and Eyal-Giladi 1971; Neff and Malacinski 1982).

It is not difficult to imagine that what is true of anatomy may also be true of physiology. Indeed, microgravity leads to a shifting of body fluids, and such shifts affect the cardiovascular system in humans. As a result, we may have an opportunity to study how the functioning of the cardiovascular system—or rather its malfunctioning—is connected with the deterioration of muscles. This of course is a matter of potential significance for the general population, especially for the elderly. Moreover, in microgravity, we no longer need many of our big muscles to support us. As a consequence, the body begins to reduce its levels of calcium and other minerals needed to strengthen them. This presents a big problem for astronauts, whose bones become weak and brittle. On the other hand, their problem may give us a chance to study the connections between bone and mineral metabolism and endocrine action. Here, the adverse reactions of astronauts to weightlessness resemble the symptoms of some diseases on Earth. Space physiology thus offers a chance to investigate the underlying mechanisms (Rambout 1981). Vestibular research, for example, may yield some insights about Meniere's disease, an affliction of the middle ear characterized by deafness and vertigo. Equivalent investigations should be carried out on Mars.

This fine-tuning of physiology to the Earth's gravity should provide a fruitful theoretical perspective to study the relationships between a variety of internal systems and cycles in the human body. Are physiological functions maximized at 1 g, or can 0.34 g suffice? This leads to questions about why the body works as it does, questions

[3]A good source for the state-of-the-art research on how gravity affects life can be found in the proceedings of the annual meetings of the IUPS Commission on Gravitational Physiology, published as supplements to *The Physiologist* (1982, 1984). See also the series of reports from Spacelab entitled "Life Sciences" (1984). For possible future experimentation see *The Fabricant Report on Life Sciences Experiments for a Space Station* (1983).

[4]See the gravitational physiology supplements to *The Physiologist* cited above. Also see Experiments on Plants Grown in Space (1984). For an assessment and long-range planning of plant gravitational research see Plant Gravitational and Space Research (1984).

that would not occur that easily otherwise. One possible answer is that gravity is used to harmonize a variety of physiological systems—gravity is like the glue that holds such systems together. Once the glue is gone, they do not quite work together. And from their failure, we learn what makes them work correctly under terrestrial conditions. Another answer is that those systems change their responses in order to adapt to the new conditions. This may also give us significant clues about their normal modes of interaction with other systems or mechanisms.

The fine-tuning to 1 g may become acute in issues of development. In microgravity, a human male excretes from 1.5 to 2 L of body fluids, with pronounced reductions in the levels of sodium and potassium. By contrast, a pregnant human female is expected, in 1 g, to show an increase of 1.5–4 L over her pregnancy, with a marked retention of sodium. Since the development of the fetus follows a strict sequence in which each event must take place within a critical period, and since the availability and composition of the body fluids are essential to the proper environment in the placenta, we can readily see that disruptions at the system level affect physiological processes at lower levels. At the present time, it would be morally impermissible to have pregnant women in microgravity.

The human body is resourceful; it may be able to compensate for the effects of microgravity in a systematic fashion even during a pregnancy. But to determine whether it can, we must resort to experiments on animals (Halstead and Pleasant 1982). Equivalent experiments on Mars are essential before giving the green light to human birth, or even human pregnancy, there.

The mere fact that many systems function optimally at 1 g provides warrant for designing experiments to determine how the timing and feedback controls of development operate. This, it seems to me, is not a matter of small importance.

"Mere" systemic effects can have profound repercussions that may extend to the cellular level. Changes in the environment of the cell lead to cellular changes in shape, in ability to move, and in internal metabolism (i.e., polarity, secretion, hormone regulation, membrane flow, and energy balance). Much has been written in this and related topics (Gravitational and Space Biology 2013). Changes that take place at the systems level in the organism, such as body fluid shifts, are then bound to affect several cellular systems, change the cellular environment, and thus affect the cells themselves.

The experiments that indicated that gravity was irrelevant at the cellular level were performed in cell cultures; they did not examine cells that formed part of the complex wholes that are the cells' normal environments.[5] It is not surprising, then, that such experiments could not expose the indirect action of gravity that starts at the systems level of the organism and works its way down into the realm of the small.

[5]For this once-standard conclusion about the space environment, see Taylor (1997). For an influential experiment on cultures of embryonic lung cells, see Montgomery Jr. et al. (1977).

11.6 Conclusion

As we have seen, the scientific exploration of Mars offers extraordinary opportunities to challenge and improve geology, climatology, and specially biology. In biology, in particular, the new knowledge could have profound implications. As a result, I have argued, we have an ethical obligation to carry out the exploration of Mars in a manner that minimally disturbs the red planet, so as to preserve such scientific treasures. Any possible large-scale colonization should wait for that scientific exploration to be completed, while greatly benefiting from its findings. We have also seen how science must become an integral part of complying with our ethical obligations to those future explorers.

References

Bylinsky, G. (1981). *Life in Darwin's universe*. Doubleday.

Cairns-Smith, A. G. (2009). An approach to a blueprint for a primitive organism. In C. H. Waddington (Ed.), *The origin of life: Towards a theoretical biology, 1* (pp. 57–66) Aldine Transaction. ISBN 978-0-202-36302-8.

Carr, M. (2004). The proof is in: Ancient water on Mars. *The Planetary Report, 24*(3), 11.

de Duve, C. (2002). *Life evolving*. Oxford University Press.

de Duve, C. (2005). *Singularities*. Cambridge University Press.

Dorminey, B. (2009). Without the Moon, would there be life on Earth? *Scientific American*, 21.

Experiments on Plants Grown in Space. (1984). *Annals of Botany, 54*(Suppl. 3).

Fabricant, J. D. (Ed.). (1983). *The fabricant report on life sciences experiments for a space station*. Galveston, Texas: University of Texas Medical Branch.

Forget, F., Wordsworth, R., Millour, E., Madeleine, J.-B., Kerber, L., Leconte, J., et al. (2013). 3D modelling of the early martian climate under a denser CO_2 atmosphere: Temperatures and CO_2 ice clouds. *Icarus, 222*, 81–99.

Gravitational and Space Research (2013). *1*(1).

Halstead, T.W., & Pleasant, L.G. (1982). *1982 Space Biology Accomplishments*. NASA Technical Memorandum 86224.

IUPS Commission on Gravitational Physiology. (1982). *The Physiologist, 25*(Suppl.).

IUPS Commission on Gravitational Physiology. (1984). *The Physiologist, 27*(Suppl.).

Jönsson, K. I., Rabbow, E., Schill, R. O., Harms-Ringdahl, M., & Rettberg, P. (2008). Tardigrades survive exposure to space in low Earth orbit. *Current Biology, 18*(17), R729–R731.

Kerr, R.A. (1997). Putative martian microbes called microscopy artifacts. *Science, 278*(5344), 1706–1707.

Kochav, S., & Eyal-Giladi, H. (1971). Bilateral symmetry in chick embryo determination by gravity. *Science, 171*, 1027.

Luef, B., Frischkorn, K. R., Wrighton, K. C., Holman, H. N., Birarda, G., Thomas, B. C., Singh, A., Williams, K. H., Siegerist, C. E., Tringe, S. G., Downing, K. H., Comolli, L. R., & Banfield J. F. (2015). Diverse uncultivated ultra-small bacterial cells in groundwater. *Nature Communications*. https://doi.org/10.1038/ncomms7372.

Martel, J., & Young, J. D. E. (2008). Purported nanobacteria in human blood as calcium carbonate nanoparticles. *Proceedings of the National Academy of Sciences, 105*(14), 5549–5554.

McKay, D. S., Gibson Jr., E. K., Thomas-Keprta, K. L., Vali, H., Romanek, C. S., Clemett, S. J., Chillier, X. D. F., Maechling, C. R., & Zare, R. N. (1996). Search for past life on Mars: Possible relic biogenic activity in martian meteorite AL84001. *Science, 273*, 924–929.

Montgomery Jr. P. O., et al. (1977). The response of single human cells to zero-gravity. In R. S. Johnston & L. F. Dietlin (Eds.), *Biomedical results from skylab*, NASA SP-377, 221.

Munevar, G. (2014). Space exploration and human survival. *Space Policy, 30,* 197–201.

National Academies of Sciences, Engineering, and Medicine. (2018). *Review and assessment of planetary protection policy development processes*. Washington, DC: The National Academies Press. https://doi.org/10.17226/25172.

Neff, A. W., & Malacinski, G. M. (1982). Reversal of early pattern formation in inverted amphibian eggs. Proceedings of the Fourth Annual Meeting of the IUPS Commission on Gravitational Physiology. *The Physiologist, 25*(Suppl.), 119.

Orosei, R., Lauro, S. E., Pettinelli, E., Cicchetti, A., Coradini, M., Cosciotti, B., Di Paolo, F., Flamini, E., Mattei, E., Pajola, M., Soldovieri, F., Cartacci, M., Cassenti, F., Frigeri, A., Giuppi, S., Martufi, R., Masdea, A., Mitri, G., Nenna, C., Noschese, R., Restano, M., & Seu, R. (2018). Radar evidence of subglacial liquid water on Mars. *Science, 361*(6401), 490–493.

Plant Gravitational and Space Research. (1984). *Report of a workshop held April 30–May 2, 1984 in Rosslyn*. Virginia: Publication of the American Society of Plant Physiology.

Rambout, P. C. (1981). The Human Element. In *A meeting with the universe: Science discoveries from the space program*. NASA, 142.

Reports, Spacelab. (1984). Life sciences. *Science, 225,* 205–234.

Science. (2007). Special section: Mars reconnaissance orbiter. Vol. 317, 1705–1719.

Siegel, S. M. (1968). Experimental Biology of ammonia-rich environments: Optical and isotopic evidence for vital activity in pennicillium in liquid ammonia-glycerol media at −40 C. *Proceedings of the National Academy of Sciences, 60*(2), 505.

Siegel, S. M. (1970). Experimental biology of extreme environments and its significance for space bioscience, 1 and 2. *Spaceflight, 12*(128–130), 256–299.

Siegel, S. M., & Spettel, T. W. (1977). Life and the outer planets: II. Enzyme activity in ammonia-water systems and other exotic media at various temperatures. *Life Science and Space Research, 15*, 76.

Siegel, B. Z., & Siegel, S. M. (1980). Further studies on the environmental capabilities of fungi: Interactions of salinity, ultraviolet irradiation, and temperature in penicillium. In R. Holmquist (Ed.), *Gospar Life Sciences and Space Research*, Vol. 8. Pergamos Press, 59.

Squyres, S. (2009). Quoted in *Space.com*. https://www.space.com/6972-steve-squyres-robot-guy-humans-mars.html.

Taylor, G. R. (1997). Cell biology experiments conducted in space. *BioScience, 27,* 102.

Toon, O. B., Pollack, J. B., Ward, W., Burns, J. A., & Bilski, K. (1980). The astronomical theory of climatic changes of Mars. *Icarus, 44,* 552–607.

Yin, A. (2012). Structural analysis of the valles marineris fault zone: Possible evidence for large-scale strike-slip faulting on Mars. *Lithosphere, 4*(4), 286–330.

Chapter 12
Ethics for an Uninhabited Planet

Erik Persson

Abstract Some authors argue that we have a moral obligation to leave Mars the way it is, even if it does not harbour any life. This claim is usually based on an assumption that Mars has intrinsic value. The problem with this concept is that different authors use it differently. In this chapter, I investigate different ways in which an uninhabited Mars is said to have intrinsic value. First, I investigate whether the planet can have moral standing. I find that this is not a plausible assumption. I then investigate different combinations of objective value and end value. I find that there is no way we can know whether an uninhabited Mars has objective end value and even if it does, this does not seem to imply any moral obligations on us. I then investigate whether an uninhabited Mars can have subjective end value. I conclude that this is very plausible. I also investigate whether an uninhabited Mars can have objective instrumental value in relation to some other, non-Mars related end value. I find also this very plausible. It is also highly plausible, however, that spreading (human or other) life to a presently uninhabited Mars can also have subjective end value, as well as objective instrumental value. I mention shortly two ways of prioritising between these values: (1) The utilitarian method of counting the number of sentient beings who entertain each value and determining the strength of the values to them. (2) Finding a compromise that allows colonisation on parts of the planet while leaving other parts untouched. These methods should be seen as examples, not as an exhaustive list. Also, I do not take a definitive stand in favour of any of the two approaches, though it seems at least prima facie that the second approach may have a better chance of actually leading to a constructive result.

12.1 Introduction

Popular media report on an almost daily basis of someone's plans to put humans on, and eventually, colonise Mars (see, e.g. Bachman 2018; Beall 2018; Brueck 2018;

E. Persson (✉)
Department of Philosophy, Lund University, Lund, Sweden
e-mail: erik.persson@fil.lu.se

Center of Theological Inquiry, Princeton, NJ, USA

K. Szocik (ed.), *The Human Factor in a Mission to Mars*,
Space and Society, https://doi.org/10.1007/978-3-030-02059-0_12

Connor 2017; Eck 2018; Griffin 2017; Mack 2017; Solon 2018; Thompson 2018). The plans can originate from a host of different sources. They include national and regional space agencies such as the US space agency (NASA), the European space agency (ESA), or the Japanese (JAXA), Chinese (CNSA) or Indian (ISRO) space agencies, but they also include commercial initiatives such as SpaceX and Mars One. The timeline is probably rather long, but one thing to remember is that technological development and business decisions (in the case of the commercial initiatives) usually happen on a much shorter time scale than for example international treaties, not to mention ethical theory. It is therefore high time to start discussing the dos and don'ts of Mars colonisation while there is still time to influence the process.

Before we start colonising, or even visiting Mars, we need to do our best to find out if there is life or not (Persson 2014). Going to Mars will be challenging whether or not there is life, but the challenges will be of different kinds, from an ethical as well as from a practical perspective.

There is an ongoing debate about what kind of ethical duties it confers on us if there turns out to be life on Mars (see, e.g. Callicott 1992; Chon-Torres 2018; Cockell 2005, 2011a, b, 2016; Green 2014; Lupisella 1997, 2000, 2009; Marshall 1993; Persson 2012, 2013, 2014, 2017; Peters 2018; Race and Randolph 2002; Randolph and McKay 2014; Rolston 1986; Smith 2009, 2016)—but what if there is no life? Will there still be ethical restrictions for what we are allowed to do on Mars even if there is no indigenous life? As long as we do not know whether there is life on Mars or not, we have to be prepared for both eventualities, and for the different challenges they pose. I have previously argued that if we encounter extraterrestrial life, the ethical restrictions that imposes on us will vary depending on what kind of life we will be dealing with (Persson 2012). In this chapter, I will explore the following question: If Mars is uninhabited, do we have a moral obligation to keep it that way?

12.2 The Intrinsic Value of an Uninhabited Mars

A common basis for arguing that we should preserve Mars in its pristine state is that a lifeless Mars has *intrinsic value* and that this, in turn, confers a moral obligation on us keep it that way. It may be because it has intrinsic value *despite* being lifeless or that it has intrinsic value *because* it is lifeless.

In order to analyse this argument, we first have to take a closer look at what seems to be the crucial term here, namely 'intrinsic value'. What does that really mean? To most people who come across it, and seemingly to most people who use the term, it seems to just mean that something has a kind of value that is very important and that places moral duties on us. For a philosopher, this is not enough, however. We like to know what, if anything, makes it so important and what that, in turn, means in terms of moral duties.

A notorious problem with the term 'Intrinsic value' is that different authors use it in different ways, and sometimes in more than one way in the same line of reasoning. Some authors have, however, tried to figure out which different concepts hide behind the term, how these concepts relate to each other, and what are their moral implications (see, e.g. O'Neil 1997; O'Neill 1992; Persson 2008; Rabinowicz and Rønnow-Rasmussen 1999).

I have previously (Persson 2008) identified five common meanings of the term 'intrinsic value', three of which I think might be useful for our analysis:

1. Value as an end rather than as a means to something else (also called 'end value', 'final value' or 'non-instrumental value').
2. Value emanating from internal properties and not from external properties (also called 'inherent value').
3. Value emanating from the phenomenon itself independently of its relations to other phenomena (also called 'non-relational value').
4. Value that exists independently of any valuer (also called 'objective value').
5. The status something has when it is the object of moral duties, that is when moral agents according to some moral system have a duty to consider its interests (also called 'moral standing' or 'moral status').

Interpretations 2 and 3 can in some types of deliberations be very important but they are not of crucial importance in this investigation. The first, fourth and fifth interpretations are very important, however, since it is typically one or more of these that proponents of instrumental value for nonliving environments seem to be thinking about. The first and forth interpretations are used for establishing moral duties in two steps: 1. Establishing the value. 2. Establishing how this value gives rise to moral duties. The fifth interpretation only needs one step since according to this interpretation, stating that an uninhabited Mars has 'intrinsic value' is just another way of saying that it endows us with moral duties. Let us, therefore, start from that end.

12.3 The Moral Standing of an Uninhabited Mars

If 'intrinsic value' just means the same as 'moral standing', we do not have to go through the first step and establish value in a real sense. Instead, we can approach the second step directly, and ask: How can we have moral duties to an uninhabited planet?

One author who argues that a nonliving Mars can have intrinsic value in this sense is Alan Marshall (1993). He does not explicitly distinguish between the different meanings of 'intrinsic value' though it seems that he is mostly talking about it in the sense we call moral standing. (In some passages, he seems to think about 'intrinsic value' more in the sense of objective value but we will get back to that later). One of his arguments for moral standing for planets and other nonliving objects builds on the successive extension of the realm of moral objects we have seen through human history from only including male members of the same community to also including women and people from other ethnic groups, to now also including animals.

According to Marshall, it is a natural development to continue this extension to also include plants, microbes and eventually abiotic objects (Marshall 1993).

This argument seems to be a positive version of what is often referred to as the fallacy of the slippery slope. Slippery slope arguments usually take the form 'Do not do A unless we also have to do B, C and so on all the way to Z'. Marshall's argument seems to take the form 'Since we have done A, B and C we should also do D, E and so on all the way through Z'. That this is a fallacy is quite easy to see. That we have said A does not in itself force us to say B. What is needed to make the argument work is some kind of mechanism that makes it inevitable that if we take the first step we also have to take the next step. On a real slippery slope, it is the slipperiness in combination with gravity that makes this happen. Once we step out on the slope, we will inevitably—because of the slipperiness and gravitational pull—slide all the way down. In our case, we would need some kind of mechanism that forces us to continue to extend the realm of moral objects once we have started, and to do it at least until we have also included abiotic objects. Marshall has not suggested such a mechanism, however, and I cannot see what it would be. The fact that we have extended the realm of moral objects several times before is not in itself such a mechanism (therein lays the fallacy). Neither can it be used as an indication that there must be one. Each time we have extended the realm of moral objects in the past, we have had good reasons for that particular extension. If we are going to extend it again in the future, we will also need good reasons to make that particular extension.

In addition to his version of the slippery slope argument, Marshall (1993) does supply some independent reasons why we should attribute moral status to abiotic objects. One argument for why we should, according to Marshall, include whole planets in the realm of moral objects, is that the environments of these planets are not static. This means according to Marshall that these planets manage to live up to many definitions of 'life'.

It is true that activeness in some sense does play a part in many attempts to define 'life'. It would be a bit of a stretch to equate the geological and meteorological activity of Mars with the kind of activity one can find in living beings, but even if we grant that activity is activity and should be handled equally, it does still not give us what we need. For one thing, being active is never in any sense seen as sufficient for being alive. More important, however, being alive cannot be sufficient for having moral standing.

Moral standing is usually analysed in terms of interests. This means, in order to have moral standing, one has to have some kind of subjective perspective from which things can go better or worse (see, e.g. Bernstein 1998; Clark 1977; de Grazia 1996; Helm 2002; Jamieson 1998; Levine 1997; O'Neil 1997; Persson 2008, 2012; Regan 2001, 2004; Singer 2009, 2011). The ability to feel pleasure and pain is often set as a minimum criterion. The reason why the possession of interests is usually assumed to be necessary for moral standing is that it seems impossible to even make sense of the idea that we have moral obligations to something that does not have any interests of its own to consider. Is it for instance in the interest of a piece of granite to be revered where it stands or is it rather in its interest to be sculptured into a statue? If the granite does not have its own perspective from which to judge this, then any judgement in the matter would have to come from the outside and we cannot reasonably claim that any duties regarding the piece of granite are duties to the granite.

When it comes to a lifeless Mars, it is very difficult indeed to imagine how a planet could have any subjective perspective to consider. One can, of course, imagine that in the future, or somewhere in another solar system, there is or will be machines that do not qualify as being alive but do have feelings. If that is or ever will be the case, I see no reason to deny these machines moral standing. This is not the situation we are talking about when it comes to present-day Mars, however. It, therefore, seems safe to say that the idea that we have moral duties to Mars as such, or to any specific lifeless object on Mars, cannot be correct.

This means that the fifth and most straightforward interpretation of 'intrinsic value' cannot be the interpretation we are looking for if we want to maintain that we have a moral obligation to preserve Mars in its pristine, lifeless, state. Maybe we will do better with the two-step approach of the first or forth interpretations?

12.4 The Objective Value of an Uninhabited Mars

Attributing objective value to something, usually means that it has a value that is independent of whether anyone values it. There is a vivid ongoing discussion about whether values can be independent of a valuer and I will not attempt to give an overview of that discussion, or for that matter contribute to it. Instead, I will just for the sake of this analysis, assume that objective value is possible. Then, the questions will be: How do we know if an uninhabited Mars has objective value and if it does, can that endow any moral obligations on us to preserve it that way?

As mentioned in the previous section, Marshall (1993) does not just believe that a lifeless Mars has moral standing. He also claims that an uninhabited Mars can have intrinsic value in the fourth sense of 'intrinsic value', namely that it has objective value. Marshall himself does not make this distinction but if we follow his reasoning, it does tend to point towards different interpretations. Some of his arguments, like those we examined in the previous section, try to directly establish moral duties. Other arguments try to establish moral duties via the establishment of objective value. Rolston (1986) is another proponent of intrinsic value for Mars as well as other planets and abiotic phenomena and who reason along similar lines, while Cockell and Honeck (2006) are more careful and just leave the door open to the possibility that Mars has intrinsic value in the sense that it has objective value.

One of Marshall's arguments goes as follows: "Although it [i.e. Mars] might seem to be a great useless hunk of red rock to us, humans could, in the view of martian rocks, be merely living organisms who are yet to attain the blissful state of satori only afforded to non-living entities." (Marshall 1993).

It is hard to know exactly what is meant by this. Marshall does not seem to think that rocks are conscious, so a literal interpretation would not do justice to the argument. A more realistic interpretation would be that not being alive can also be a state that has value in itself, even though it cannot be experienced. Neither can it be fairly judged by us living beings.

Based on Marshall's text, it is impossible to know for sure if this is the correct interpretation but it is the most reasonable I have managed to come up with. Unfortunately, it is still not convincing enough, however. The big hole in the argument seems to be in the leap from the fact that we cannot know that a state of nonliving does not have objective value, to accepting that we should assume that it does and let that assumption determine how we should act.

This actually points straight at two major, and seemingly insurmountable problems with basing moral obligations on assertions of objective value. There does not seem to be any way of determining what has objective value or why. Therefore, we have no good answer to how objective values can entail moral duties, or what these duties would be.

Rolston (1986) provides many good reasons to value geological features, including lifeless planets. He writes "… they are recognizably different from their backgrounds and surroundings. They may have striking particularity, symmetry, harmony, grace, spatiotemporal unity and continuity, historical identity, story, even though they are also diffuse, partial, broken." There are several passages similar to this one where he depicts non-biotic nature in very poetic terms. He also—like Marshall—points out that the abiotic nature is active in its own right, but while Marshall uses this as an argument in favour of moral standing, Rolston argues that this is another reason to value it (Rolston 1986). It is easy, when reading this, to see that there are good reasons to attach value to these phenomena. It is not easy, however, to see how this value is independent of a valuer. Rolston does a very good job in showing how the phenomena he describes fit in our normal sense of beauty, awe, etcetera, but how would the features he describes using human language fare if there were no humans with our particular sense of value? It seems impossible to know the answer to that question and therefore also to answer the general question what has objective value and why. This also makes it difficult to answer the corollary question of why this would lay any moral duties on us.

Marshall's other main argument in favour of objective value for an uninhabited Mars is inspired by Rolston (1986). They both agree that we should appreciate nonliving worlds for what they are instead of 'depreciating' them for not harbouring life (Marshall 1993). He finishes by saying that "In reality they have not 'failed' to be anything; they have achieved being what they are. We must not consider Mars or any other celestial body to be unlucky just because it does not support life. Indeed, even in the absence of an indigenous lifeform, Mars possesses its own uniqueness and diversity which are worthy of respect." (Marshall 1993).

Like with Rolston's poetic description above, Marshall's argument presents very good reasons for us sentient beings with emotions that answer to properties like those Marshall mentions, to value Mars as an end in itself, even without life. The question that remains to be answered is also the same as with Rolston: how do we know that Mars has value even if we were to fail to appreciate it, and if it does, why would that entail any moral duties on us?

Marshall (1993) also argues that "It is evident that geologists can admire and wish to preserve the pristine nature of geological structures, just as a biologist would wish to preserve a living forest or coral reef." This is, again, a good example of a subjective value. This does not have to mean that this argument, or the previous ones, are not useful in our investigation (we will get back to the question of subjective values in the next section). It just means, it is not a good argument for objective value.

The question of whether any objective value of an uninhabited Mars confers any moral duties on us to preserve it that way is interesting in its own right and will, of course, be very relevant if anyone manages to figure out a way of determining what has objective value. It is usually assumed by those who refer to objective values that if something has objective value, then we have to be morally obligated to protect it. I think, however, that there are good reasons to question that assumption.

In the previous section, we noted that interests seem to be necessary for moral standing. This means that no matter how much objective value something has if it does not have any interests of its own and does not matter to anyone with interests (that is if it does not have subjective value), it cannot generate any moral duties. This, in turn, means that the possible existence of objective value on Mars or anywhere else is ethically irrelevant.

Let us try to imagine what it would mean by imagining something that is typically not valued by valuers today and imagine that it instead has objective value and that this value generates moral duties. Let us say, for instance, that putting things on top of other things (courtesy of Monty Python) has such value. It may seem like a silly example but remember that we do not know what has objective value, and we need to assume that it is different from the more well-known subjective values, so we will not confuse the matter by subconsciously projecting our subjective values on the phenomenon we are investigating. According to the argument, the objective value of putting things on top of other things would generate a moral duty to put things on top of other things. Does this seem reasonable? I suppose a possible explanation to why it does not is that we as valuers are biased in favour of things we value today. This is certainly possible, even plausible, but I must still confess that if we sometime in the future manage to determine that some things do have objective value, and realise that these things differ from what has subjective value to sentient beings (that is to beings who can actually suffer or rejoice based on whether the things they value are promoted/preserved or not) and we at least in some instances have to choose between protecting or promoting objective values that do not matter to any sentient being, and protecting or promoting some competing subjective value, the preservation or promotion of which, brings joy or mitigates suffering among sentient beings, then I for one would go for the subjective value every time. I would, in fact, consider it quite immoral to do otherwise.

Let us not give up yet, however. I mentioned Monty Python's example of putting things on top of other things because it is a clear example of something that is not valuable from the subjective perspective of most readers, but maybe there are better examples?

If someone decided to strip mine Mount Everest, the highest mountain on Earth, it would probably evoke quite a lot of protests. This can of course easily be explained by the subjective value it has in the minds of the protesters, but what about Olympus Mons? Olympus Mons is the largest mountain on Mars as well as in our solar system. If we assume, as we do in this chapter, that Mars is lifeless, then does not that mean that since no one is there to value it, it must lack subjective value? Does it not at the same time, seem rather odd that Mount Everest deserves to be protected because of its exceptional statute, while a mountain that is three times as high on another planet does not?

This would seem like a good case for claiming that some objects do deserve respect even though they are not valued by any sentient being, and they are not sentient beings themselves. We have to remember, however, that even though there is no one (we assume) on Mars to value it, we are still talking about it being threatened by strip-mining. This can only be the case because we humans on Earth know about this mountain. We know that it exists, we know how high it is, we know at least a bit about its geological history, and we can in fact even see it through telescopes and in photographs taken through telescopes and by instruments in orbit around the planet. Saying that it cannot have subjective value because Mars is uninhabited is therefore not completely true. It can still be valued subjectively by us Earthlings.

So, what about an object on some planet around another star that we can still not observe and do not know anything about. Cannot that object have value in such a way as to generate moral duties? If strip-mining Mount Everest and Olympus Mons would be bad, would it not be even worse if something happens to another mountain even higher and more majestic than Olympus Mons on another planet around another star?

If we agree that it would, it is because we as sentient beings have formed the concept of a mountain that is higher and more majestic than Olympus Mons and because we have also formed an abstract principle of high and majestic mountains having value, which still makes it highly doubtful that this example can be used as evidence of objective value. There is also another problem with this example. Even if we did assume that an extrasolar mountain that is higher than any mountain in our solar system would have objective value, and even if it was threatened by some strip-mining project, there is no way we can do anything about it. We can therefore not have a moral obligation to protect it (as long as we accept that ought implies can). This is not just for the contingent reason that it will be very far away, and we have yet to invent warp-drive (courtesy of Star Trek). In this particular case, there is an even more compelling reason, namely that the very basis for this thought experiment was that we are not aware of this mountain. If we were, the basis for concluding that the value of the mountain is not subjective would fail for the same reason as with Olympus Mons.

12.5 The Subjective End Value of a Pristine Mars

The first interpretation of 'intrinsic value' in the list says that an uninhabited Mars can have value as an end in itself. This concept is compatible with the idea of objective value as well as with an idea of subjective value. When we discussed the idea that an uninhabited Mars has objective value, it was assumed that we were talking about end value. The concept of end value is, however, equally compatible with the idea that an uninhabited Mars has subjective value. In that case, the claim is that there are people who simply value Mars the way it is even if (or because) there is no life and that the rest of us have an obligation to take that into consideration when deciding what to do with Mars.

Here we need to ask two questions: 1. Is it reasonable to assume that an uninhabited Mars can have value as an end to sentient beings on another planet (that is Earth)? 2. Can such values if they exist generate any moral duties?

We have in fact already established a positive answer to the first question through our discussion about Olympus Mons above. Both Rolston's (1986) and Marshall's (1993) arguments for why an uninhabited Mars has objective value, were found to fit much better with the notion that they have subjective value, at least to the authors, but probably also for others who share the emotions the authors expressed in the text.

When it comes to end value, it seems, in fact, quite irrelevant whether the valuer and the valued are residing on the same planet or not. End value is the kind of value something has because it is valued in its own right independently of whether it can be used to achieve some other value. Even if no human will ever visit Olympus Mons and even if we will never be able to use it for anything, mining, tourism, sport, or anything else, it can still have value to us by just standing there and be tall and majestic. The same must be true for other features on Mars, as well as for the planet itself.

Through the reasoning in the previous section we also, partly, answered the second question, whether subjective end values can generate moral duties. We established in that section that things that have value subjectively to sentient beings are the only things that can generate moral duties. This also means that if we accept that there is such a thing as moral duties, these duties have to be generated by things having subjective value to sentient beings.

Does this mean that since some people attach end value to Mars in its uninhabited state, we have to leave the planet in that state? The answer to that question is no. Just as there are people who value Mars in a lifeless state, there are those who find great value in making Mars come alive and there is no reason why not this too is a legitimate end value.

The obvious question would then be, what are we morally obliged to do, knowing that people value completely opposite states? Does not this mean that grounding moral duties in subjective values is in fact impossible?

If it was the case that as soon as someone subjectively values something for its own sake, then we all have a duty to preserve or promote that thing, then we would end up in an impossible situation in all cases where people have opposing values. Luckily, there is no reason to go as far as saying that all subjective end values generate absolute duties to promote or preserve them. The reasonable approach is instead to acknowledge that we are morally obliged to consider these values and attach due weight to them when we make decisions.

So, how are we to do that? If it was possible to base moral duties on objective value and if it was possible to identify objective values, we would not have this problem. We would 'just' have to ask ourselves, who is right? Unfortunately, as we saw in the previous section, this is not an option.

The option that is advocated by utilitarianism tells us that even though the values at stake are subjective, it is still possible to find a true, objective, answer to the question of what our moral duty is, namely by counting the number of people holding respective value and determining the strength of the values to those who hold them. It is usually practically impossible to do this with a particularly high degree of precision, but it may be possible to provide reasonable approximative answers.

Another way of dealing with the dilemma would be to acknowledge that there is no objectively true answer and sit down and talk with the aim of finding a compromise that those with an interest in the matter at least can live with even if it is not ideal. This approach may lack the theoretical attractiveness of finding objectively true answers, either to what is of value or to how to transform subjective values into objective duties. Contrary to these approaches, however, it may actually work, which I think, may be at least as attractive. One outcome of such an approach could be a compromise in the form of the establishment of a set of protected areas (suggested by among others Cockell and Honeck 2004, 2006 and Rolston 1986).

12.6 The Instrumental Value of a Pristine Mars

We noted above that if it was possible to identify and base moral duties on objective values, it would be much easier to handle value conflicts. We also found that the prospect of doing that seems bleak, but maybe we, despite our efforts above, have not totally exhausted the possibilities? Let us give objective values one more try.

Let us start with another thought experiment, this time with a very simple and down to Earth one. Let us imagine a 2-year old human who really hates eating. She is clearly a sentient being and she clearly assigns a negative value to eating. To her, eating has a very strict negative subjective value. As an adult, I know that eating is, in fact, necessary for one's survival, whether one likes it or not, and as a responsible parent, I use every trick in the book to make the 2-year old eat, despite her very clear and, very clearly stated, preference to the contrary.

Is not this a good illustration of a case of something, namely eating, that has objective value in the sense that it has value whether one likes it or not, and is not life, in fact, full of cases like this? Some tools are objectively, measurably better for performing a certain function than other tools whether we like them better or not. Some football players are just better than others at scoring and this can be found out objectively just by looking at the scoreboard.

We can with ease find an almost unlimited number of examples like this and they are all good solid illustrations of the fact that there are, indeed, identifiable objective values—in a way. We can also with ease find examples of how these cases generate moral duties. If someone is sick and you are a physician, it seems obvious that you have a moral duty to use methods and equipment that is objectively better at curing this disease, at least as long as they are available and as long as it does not conflict with other moral duties.

It thus seems that we have, finally, found a way in which values can be objective and in which these values do create moral duties. There is one thing we have to notice, however. In addition to being objectively determinable, all the values mentioned in this section so far also have one other thing in common. They are examples of phenomena that are only valuable in so far as they can produce a result that is subjectively valuable. Eating and curing diseases are valuable only to the extent that we value life, or the things continued living can give. The football player is good at scoring, which is only good if we root for her team. This may seem too obvious to point out, but it is easily overlooked, and it is important. The values we discussed in previous sections were end values. All the values we have referred to here, are instrumental values. We have thus left the realm of so-called 'intrinsic values'. This does not necessarily have to be a problem, but it does have some repercussions that need to be pointed out. Strictly speaking, only end values are in some sense 'real' values. Stating that something has instrumental value is just a convenient way of stating a belief (that may or may not be true) that something tends to promote or preserve something else that has value. This, in turn, has the practical implication that we are only morally obligated to promote or preserve instrumental values if they are important promoters or preservers of something else that we are morally obligated to preserve or promote because it has end value to sentient beings.

How does this help us understand our moral obligations regarding an uninhabited Mars? It does by clarifying that an uninhabited Mars may not only have end value to sentient beings whose interests we are morally obligated to consider. Keeping Mars lifeless might also have instrumental value in the sense that it helps us promote or protect something else that has end value to sentient beings, and that do not necessarily have to do with Mars.

How then can keeping Mars uninhabited be instrumentally valuable to us earthlings? Billings (2006) points out that putting humans on Mars will affect human society and culture on Earth. In the same paper, she warns about the attitude taken for granted that we should spread western consumerism into space. She also questions the American frontier metaphor that she claims is not even historically valid on Earth though it seems to be a very strong metaphor, frequently used to support a rather aggressive form of space colonialism (Billings 2006). Following this line of thought, we can see that abstaining from treating Mars as a frontier to be conquered or colonised, or as a resource to be exploited in order to uphold a consumerist lifestyle, could be seen as a moral duty based on the value of not upholding a flawed lifestyle and a set of flawed attitudes that have a negative instrumental value to most of us (whether we realise it or not) by destroying things with end value on our own planet. If abstaining from colonising Mars can help curbing these habits on Earth is an empirical question that is outside of the scope of this chapter, though it does not seem completely implausible.

A rather speculative twist to the question of instrumental value is the idea that uninhabited worlds should be left alone for the sake of future life that may evolve on the planet. Green (2014) points out that Immanuel Kant seems to have been of the opinion that we do have such a duty. Green does not believe that this is a strong argument for leaving Mars alone, however, since he does not believe that the prospects for life to evolve on Mars without help from Earth are very bright. As Green also points out, Kant was a proponent of the view that only intelligent life has moral status. Since the probability that intelligent life will evolve on Mars ought reasonably to be even smaller than the probability that any life will evolve on the planet, this makes it even less meaningful from a Kantian perspective to impose a moral duty on us to preserve the planet in its pristine state. A more plausible account of what it takes to have moral status, that draws the line at sentience rather than intelligence, would, of course, increase the probabilities somewhat, but the actual probability would still be negligible.

One interesting twist that seems to present itself here is whether Kant's view cannot be used more successfully as an argument in favour of increasing the probability that life will evolve on the planet by altering the conditions of the planet, so they become more similar to those on Earth at the time life first appeared here—or even seed the planet with Earth life.

Green does not discuss this question and since it falls outside the scope of this chapter, I will not delve into it either, other than by noting that Kantian ethics normally consider negative duties to be much stronger than positive duties, which in this case reasonably would favour leaving the planet alone over altering or seeding it.

One of Marshall's (1993) arguments for the objective value of Mars is 'Even if a planet appears undynamic or dead it may preserve in its rocks and minerals things which represent millions, or even billions of years of past dynamic processes. It may also preserve past histories of life in the form of fossils.' Marshall means this as an example of an objective end value. As noticed above, however, we have no way of assessing whether representing billions of years of dynamic processes really is an objective end value, though it is not unreasonable to assume that it does represent a subjective end value to some. It is, however, also an excellent example of instrumental value. By preserving traces of geological dynamic processes, a lifeless Mars can be very useful for science. On Earth, the geological processes are intertwined by biological processes. If Mars is uninhabited, it will provide opportunities to study pure geological processes, which means an uninhabited Mars has instrumental value in relation to science and as a result of that, to knowledge. Both science and the resulting knowledge are examples of things that can have both end value and instrumental value in relation to other values that can be promoted or preserved by the help of knowledge.

On the other hand, if we choose to seed Mars with Earth life or terraform and colonise Mars, science will have the opportunity to see how life, and eventually society, will evolve on another world. This too will create knowledge that may have end value and/or instrumental value. Instrumental values are thus objective in the sense that some are identifiably better at promoting and preserving end values than others no matter what we happen to think about them. On the other hand, something's status as and degree of instrumental value depends ultimately on the end values they preserve or promote. This, in turn, means that if, and to what strength, they generate any moral obligations, also depends on the subjective end value they ultimately promote or protect.

This means that this last attempt to find objective value that can help us determine once and for all whether we have a moral duty to preserve Mars in its pristine (presumably) lifeless state did not get us all the way but provided instead another class of values that we have to consider.

Whether the examples we have seen in this section, or any other instrumental value an uninhabited Mars may have, provide a moral obligation to preserve Mars in its pristine, lifeless state depends, as we have seen, ultimately on the subjective end values that these instrumental values preserve or promote, as well as how important they are for preserving or promoting these end values. It depends also on how we prioritise or compromise in the face of opposing end values.

12.7 Conclusions

If Mars does not harbour any indigenous life, do we have any moral obligations to keep it that way?

We have not found any unambiguous or straightforward answer to the question, but we have managed to straighten out several questions surrounding this big question. One thing we have found is that the planet Mars does not in itself have any 'say' in this matter. Non-sentient entities, like planets, whether they harbour life or not, do not have a moral status. We, therefore, do not have any moral obligations to the planet itself to keep it in its pristine, lifeless state.

We also found that we cannot settle the matter by referring to objective end values. Even if they exist, there is no way of identifying them in a way that clearly set them apart from subjective values. It is also not possible to base any moral obligations on objective end values. Since we can only have moral obligations towards sentient beings, we can only have moral obligations regarding things that are valued by sentient beings, that is, regarding subjective end values.

We did find some plausible ways in which an uninhabited Mars may have end value to some sentient beings (most plausibly humans on planet Earth). This does not necessarily mean that we have a moral obligation to leave Mars in that state, however. We know that some people think that spreading life to Mars would have end value. How to handle this value conflict lays beyond this investigation. We can merely notice that there may be different ways of doing that. A utilitarian would go about this task by trying to determine the number of people on each side and to assess the relative strength of the competing values to the valuers. Others have suggested a compromise in the form of leaving some parts of Mars in its pristine state while colonising other parts.

There is one type of objective values that are relevant to our investigation, namely instrumental values. They are objective in the sense that it is at least in principle possible to objectively determine whether something is better than something else at promoting or preserving a certain end value. Ultimately, however, whether something has instrumental value and how much, depends on the subjective end value it promotes or protects.

We did manage to identify some plausible ways in which it may have instrumental value to leave Mars uninhabited. Whether that is enough to generate any definitive moral obligation to leave Mars that way, depends ultimately on how we prioritise between the end values these instrumental values promote or protect and other end values including the positive or negative valuation of an uninhabited Mars in itself.

Acknowledgements The project was funded by The National Aeronautics and Space Administration through grant NNX14AR81G.

References

Bachman, J. (2018). Setting a date for landing humans on Mars *Bloomberg*. https://www.bloomberg.com/news/articles/2018-05-08/can-humans-get-to-mars-in-12-years. Accessed 26 Sept 2018.
Beall, A. (2018). If Elon Musk is to colonise Mars, he'll need to recruit a crew of genetically-modified humans *Wired* 03/02/2018. https://www.wired.co.uk/article/elon-musk-mars-colony-spacex-radiation-genetically-modified-humans. Accessed 26 Sept 2018.
Bernstein, M. (1998). Well-being. *American Philosophical Quarterly, 35*, 29–55.
Billings, L. (2006). How shall we live in space? *Culture, Law and Ethics in Spacefaring Society Space Policy, 22*, 249–255.
Brueck, H. (2018). NASA wants to send humans to Mars in the 2030s—here's the step-by-step timeline *Business Insider* 01/03/2018. https://www.businessinsider.com/when-will-we-colonize-mars-nasa-timeline-2030s-2018-2?r=US&IR=T&IR=T. Accessed 26 Sept 2018.
Callicott, J. B. (1992). Moral consider ability and extraterrestrial life. In E. C. Hargrove (Ed.), *The animal rights/environmental ethics debate* (pp. 137–150). Albany: State University of New York Press.
Chon-Torres, O. A. (2018). Disciplinary nature of astrobiology and astrobioethic's epistemic foundations. *International Journal of Astrobiology*, 1–8.
Clark, S. R. L. (1977). *The moral status of animals*. Oxford: Clarendon Press.
Cockell, C. S. (2005). The value of microorganisms. *Environmental Ethics, 27*, 375–390.
Cockell, C. S. (2011a). Ethics and extraterrestrial life. In N. -L. Remuss, K. -U. Schrogl, J.-C. Worms, & U. Landfester (Eds.) *Humans in outer space—interdisciplinary perspectives, studies in space policy* (vol. 5, pp 80–101).
Cockell, C. S. (2011b). Microbial rights? *EMBO Reports, 12*, 181.

Cockell, C. S. (2016). The ethical status of microbial life on earth and elsewhere: In defence of intrinsic value. In J. S. J. Schwartz, & T. Milligan (Eds.), *The ethics of space exploration* (pp. 167–179). Cham: Springer.
Cockell, C., Horneck, G. (2004). A planetary park system for Mars. *Space Policy 20*, 291–295.
Cockell, C. S., Honeck, G. (2006). Planetary parks—formulating a wilderness policy for planetary bodies. *Space Policy 22*, 256–261.
Connor, N. (2017). Beijing eyes 'robotic and human settlement' on Mars with ambitious 2020 mission. *The Telegraph*. Published 21/09/2017. http://www.telegraph.co.uk/news/2017/09/21/plans-2020-chinese-mars-probe-explore-possibility-human-settlement/. Accessed 23 Sept 2017.
de Grazia, D. (1996). *Taking animals seriously*. Cambridge, UK: Cambridge University Press.
Eck, A. (2018). Can we colonize Mars? *Nova next*. http://www.pbs.org/wgbh/nova/next/space/want-to-live-in-a-treehouse-on-mars/. Accessed 26 Sept 2018.
Green, B. P. (2014). Ethical approaches to astrobiology and space exploration: Comparing kant. *Mill, and Aristotle Ethics: Contemporary Issues, 2*(1), 29–44.
Griffin, A. (2017). SpaceX will colonise Mars in the next few years, says boss Elon Musk. *Independent*. https://www.independent.co.uk/life-style/gadgets-and-tech/news/spacex-elon-musk-mars-colony-moon-space-rocket-falcon-nasa-australia-a7973546.html. Accessed 26 Sept 2018.
Helm, B. W. (2002). Felt evaluations: A theory of pleasure and pain. *American Philosophical Quarterly, 39,* 13–30.
Jamieson, D. (1998). Animal liberation is an environmental ethic. *Environmental Values, 7,* 41–57.
Levine, J. (1997). Recent work on consciousness. *American Philosophical Quarterly, 34,* 379–404.
Lupisella, M. (1997). *The rights of Martians space policy 13*(2), 89–94.
Lupisella, M. (2000). *Humans and Martians earth space review 9*(1), 50–60.
Lupisella, M. (2009). The search for extraterrestrial life: Epistemology, ethics, and worldviews. In C. M. Bertka (Ed.), *Exploring the origin, extent, and future of life: Philosophical, ethical, and theological perspectives*. Cambridge: Cambridge University Press.
Mack, E. (2017). Elon musk has an update on his Mars colony plans. *CNET*. Published 20/09/2017. https://www.cnet.com/news/elon-musk-spacex-mars-colony-nasa-iac-2017-adelaide/. Accessed 22 Sept 2017.
Marshall, A. (1993). Ethics and the extraterrestrial environment. *Journal of Applied Philosophy, 10*(2), 227–236.
O'Neill, J. (1992). The varieties of intrinsic value. *The Monist, 75,* 119–137.
O'Neil, R. (1997). Intrinsic value, moral standing, and species. *Environmental Ethics, 19,* 45–52.
Persson, E. (2008). *What is wrong with extinction?*. Lund: Lund University.
Persson, E. (2012). The moral standing of extraterrestrial life astrobiology *12*, 976–984.
Persson, E. (2013). Philosophical aspects of astrobiology. In D. Dunér, J. Pathermore, E. Persson, & G. Holmberg (Eds.), *The history and philosophy of astrobiology* (pp. 29–48). Cambridge: Cambridge Scholars.
Persson, E. (2014). What does it take to establish that a world is uninhabited prior to exploitation?–A question of ethics as well as science. *Challenges 5*, 224–238.
Persson, E. (2017). Ethics and the potential conflicts between astrobiology, planetary protection and commercial use of space. *Challenges 8*(1), 12.
Peters, T. (2018). Does extraterrestrial life have intrinsic value? An exploration in responsibility ethics. *International Journal of Astrobiology*, 1–1.
Rabinowicz, W., & Rønnow-Rasmussen, T. (1999). A distinction in value: Intrinsic and for its own sake. *Proceedings of the Aristotelian Society 20*, 20–33.
Race, M. S., & Randolph, R. O. (2002). The need for operating guidelines and a decision making framework applicable to the discovery of non-intelligent extraterrestrial life. *Advances in Space Research, 30*(6), 1583–1591.
Randolph, R. O., & McKay, C. P. (2014). Protecting and expanding the richness and diversity of life, an ethic for astrobiology research and space exploration. *International Journal of Astrobiology 13*(1), 28–34.
Regan, T. (2001). *Defending animal rights*. Urbana, IL: University of Illinois Press.

Regan, T. (2004). *The case for animal rights*. Berkeley: University of California Press.
Rolston, H. I. (1986). The preservation of natural value in the solar system. In E. C. Hargrove (Ed.), *Beyond spaceship earth* (pp. 140–182). San Francisco: Sierra Club Books.
Singer, P. (2009). *Animal liberation*. New York: Harper Collins.
Singer, P. (2011). *Practical ethics*. Cambridge, UK: Cambridge University Press.
Smith, K. C. (2009). The trouble with intrinsic value: An ethical primer for astrobiology. In C. M. Bertka (Ed.), *Exploring the origin, extent, and future of life: Philosophical, ethical, and theological perspectives*. Cambridge: Cambridge University Press.
Smith, K. C. (2016). The curious case of the martian microbes: Mariomania, intrinsic value and the prime directie. In J. S. J. Schwartz, & T. Milligan (Eds.), *The ethics of space exploration* (pp. 195–208). Cham: Springer.
Solon, O. (2018). Elon musk: We must colonise Mars to preserve our species in a third world war. *The Guardian* 11/03/2018. https://www.theguardian.com/technology/2018/mar/11/elon-musk-colonise-mars-third-world-war. Accessed 26 Sept 2018.
Thompson (2018) Is colonizing Mars the most important project in human history? T*he Atlantic*. https://www.theatlantic.com/technology/archive/2018/06/could-colonizing-mars-be-the-most-important-project-in-human-history/564041/. Accessed 26 Sept 2018.

Chapter 13
Religion for a Spatial Colony: Raising the Right Questions

Lluis Oviedo

Abstract The possible scenario of a future human settlement on Mars invites reflection regarding the hypothetical conditions within which one could explore religion's possible role. The issue is what might religion's role be living a such a remote place and how religion might help to cope with extreme circumstances. The first question has to do with religious faith. The second has to do with what kind of religion or religious style would be more fitting in that highly unpredictable and threatening context. And third, what functions could religious beliefs and practices provide. The issue is serious in that it concerns the future of religious faith and its possible development and adaptation to very new contexts beyond the confines of the earth.

Among the many details when planning a mission to Mars and to establish a human settlement in the Red Planet, religion is not the one that has high priority for the engineers and specialists working on such project.

Obviously, the technical issues come first for such an endeavour, and this is not surprising. How to survive in such an inhospitable planet is the first concern. Once these technical problems are solved, we can focus on other less pressing issues that could contribute to the success or failure of the endeavour, like how to organize such a community and what rules or forms of governance to adopt; which normative systems to implement; how to keep astronauts entertained or to cope with boredom; and the most convenient emotional and sexual behaviour. Finally, other issues—in some sense called the 'ultimate'—could be considered, such as how will people have a sense of meaning and purpose to their lives; how will they keep and enforce high moral standards; and, finally, whether religious beliefs and practices like prayer, meditation or rituals would make any sense in such a strange and alien context. Being so far away from the settings that render such practices and rituals meaningful, the perceived oddness of a de-contextualized religious practice would have to be addressed. In any case, very few studies on a Mars mission and station mention such 'spiritual' or 'ultimate' issues, whose 'ultimacy' would mean that they come last in the long list of concerns for the mission designers. Even those concerned with ethical

L. Oviedo (✉)
Theology Department, Antonianum University, Rome, Italy
e-mail: loviedo@antonianum.eu

K. Szocik (ed.), *The Human Factor in a Mission to Mars*,
Space and Society, https://doi.org/10.1007/978-3-030-02059-0_13

issues care little about religious or spiritual questions (Nair et al. 2008; an exception being Szocik 2017).

However, things could be different. A Mars expedition in which we could not ensure some good degree of meaning or purpose, or in which motivations and morale were not high, could be associated with a complete failure. These issues would be related to motivations and psychological mood. They could mean a lot when trying to design an expedition in which the 'human factor' plays a key role, a level similar to the technical issues that might ensure minimal survival conditions, or a favourable, relatively friendly, environment in which that space settlement could thrive.

If we want to focus on religious or spiritual aspects in a Mars mission and colony, then we need to consider that special environment, both natural and social, and figure out what role religion could play for the well-being of that settlement. And then, we would probably need to find out which religious form or style could be more fitting, since there are many earthly religions. Perhaps, some combination of existing religious forms would offer better chances of coping with the prospective needs.

The proposed exercise invites to follow some methodological steps and to establish some previous conditions. For instance, it is very debatable in our own planet and most advanced societies, to what extent religion still plays a significant role, especially for highly formed and strong people, like those type of persons who would be selected for such a mission. Then, we can just imagine the extreme conditions for those who would live on the Red Planet. A religious or spiritual search might be expected, but the remoteness from earth and earthly religions might make that experience bizarre or weird.

The hypothetical scenario of a Mars settlement presents the opportunity for a sort of 'mental experiment' to better discern religion's functions and its relationship within a context, its evolution, as well as its potentially adaptive features. In that sense, imagining such a scenario, its psychosocial and environmental conditions could assist in assessing to what extent a religious structure—cognitive, behavioural and social or cultural—could be instrumental in coping with extreme conditions, and to what extent religion depends from context and psychological circumstances, instead of being a dimension on its own. Or, in other words, to what extent could religion be thought of as an independent variable. This is not a secondary issue: what is at stake is the way religions arise and evolve, depending on historical and environmental factors, which often assume a cultural or social character. Evolutionary concepts, such as selection, create forms that have more fitness. The 'mental experiment' is such as to test recent theories about complex relationships between religion, society and culture.

Considering the mentioned difficulties and our interests and aims, some previous experiences will work as a guide. Indeed, a good amount of data is available about missions by actual astronauts spending long periods in orbit's spatial stations, confined and relatively isolated. Relevant information is provided by castaways living in isolated and harsh environments on earth and struggling to survive. Their religious needs and experiences could be useful for the present mental experiment. In any case, we need better information about the psychological conditions that can be foreseen in such a severe context, to delineate later the possible role religion can play and which religious format could better fill the gaps.

Following the aforementioned worries and interests, the present chapter will start describing the expected conditions that could justify religious assistance, using available published studies. Second, it will move into more social and psychological studies about religious functions in advanced social and intellectual conditions, trying to find eventual connections with the astronauts' own minds and feelings. The third step will try to figure out which kind of religious beliefs and practices would fit better in a very contingent and unpredictable panorama, that of the crew. The last point raises additional questions regarding the feasibility of a designed or tailor-made religious form that could be functional for the needs of the astronauts in that extreme conditions.

13.1 Expected Psychological and Social Conditions on a Mars Mission and Settlement

We are still at a very early stage regarding a Mars mission, and considering the current scientific and technical studies, serious doubts still threaten the feasibility of such a project (Do et al. 2016). Then, some concern has been expressed regarding the current way to sell the Mars enterprise to the general public, and the unjustified expectations and illusions such a project awake (Slobodian 2015). However, many studies have been published trying to figure out the conditions—psychological, social and cultural—in which a crew would have to travel and stay on the Red Planet. It is useful to review this material and look in more depth to some of them. The aim is to describe the difficulties and challenges that a crew travelling to Mars and staying an extended period could have; in this way, I expect we can imagine how religious beliefs and practises could work.

The sources for these studies are mainly the collected data from astronauts living a long time in spatial orbital stations; winter sojourns in Antarctica in extreme and isolated settings; and the recent experiments imitating living standards in a hypothetic Mars habitat in similar desert environment, and during a long enough duration (Solcova et al. 2016).

Several studies have tried to describe the psychological problems that are more habitual for astronauts in remote locations and confinements. Nick Kanas is the author who probably has worked more on these issues and has been more active in publishing his outcomes. In a scientific comment to a fictional novel he has written on a Mars mission (Kanas 2014), he enumerates in a table the list of 'stressors'—as he calls them—that would arise in such difficult contexts. The list contains 24 items; some are worth mentioning: occupational issues, or how to occupy crew's time; relational or social ties with other crew members; communication issues in trying to exchange with Earth people and find external support; and isolation and distance increasing a sense of rootlessness and homesickness. Other stressors probably can arise if things worsen: despair or meaninglessness; tensions and conflicts between members, and even pathological moods like depression and dysphoria (Kanas et al. 2009; Kanas 2011).

Kanas is not the only one who pays attention to these problems; some other authors have tried to describe them and even to propose ways to cope with them. For instance, Dietrich Manzey (2004) distinguished between other similar living conditions and those we can imagine in a Mars mission, to show the differences among them. That comparison poses 'new psychological challenges' (Manzey 2004, 783) and argues for more attention to the implementation of strategies or procedures for dealing with those big risks. The author analyses these differences in specific fields: individual behaviour and performance, especially 'possible effects on mood, morale, and motivation of crew members'; and crew interactions. In the next step, once an assessment is made, 'psychological countermeasures' need to be designed. Again, considering what works within the current space stations, some current means of intervention will not be available, and communication delays—due to a much longer distance—will make it more difficult for psychological support by professionals interacting with astronauts, as well as support from family and friends. Communications would be reduced to emailing, and even email could be liable to periods of blackout. Current research is trying hard to fix these problems and the expectation is that communications could be improved in the near future. A solution to these new challenges is to focus more on selection and training; looking for candidates more psychologically resistant to isolation and reduced external communication and to ensure that there will be better interactions among crew members.

Manzey adds some other psychological problems shading that mission, like the so-called 'Earth-out-of-view' phenomenon, something never before experienced but quite threatening, with unknown but suspected negative effects. In the author's own words we are before a 'mysterious contingency factor': 'Thus it represents a psychological risk at least of the first mission to Mars which, in principle, will be indeterminable and uncontrollable' (Manzey 2004, 787). This loss of the usual context for human experience and action could trigger a series of negative consequences, anxiety and values crisis, as far as those which are linked to earth and its social and cultural niches. Religion or spirituality does not belong in this report as one of the 'countermeasures' repertoire aimed at coping with the described dangers and problems. Indeed, in this study nothing invited us to think on possible 'religious coping' strategies, something that has drawn scarce attention in some later approaches to the Mars mission.

Other studies have highlighted evidences about crew relationships deteriorating over time (Sandal and Bye 2015). However, several published reports focus less on problems and more on solutions, or on what could become inspiring and provide some orientation about how to proceed to address the described challenges. The examples include proposals on how isolation experiences that have been done in Japan can assist in designing coping strategies in outer space (Tachibana et al. 2017); a study on how cultural ethology could help to figure out adaptive behaviours in those risky and undetermined new contexts (Tafforin and Giner 2017); and which coping strategies have been more successful in space missions, and which could suggest similar tactics when planning a Mars settlement (Suedfeld et al. 2009). Indeed, this last author suggests that some transcendence experiences could work in some way and provide better coping resources, when showing that spatial missions report positive experiences for many participants, and not just stress (Suedfeld 2005, 2006). Highly relevant in this section is a study by Goemaere and colleagues (2016) that suggest a psychological approach—the self-determination theory—able to provide both, diagnosis about stressors, and countermeasures able to restore positive mood and to enhance the morale and group climate. This is just theoretical and possibly it suggests a strategy that would clearly compete with religious coping, or perhaps be complementary to that coping strategy. Furthermore, ethical concerns have their own place in all the debate on Mars missions (Nair et al. 2008).

Physical, psychological and social conditions in a hypothetical Mars station point to an extremely precarious situation, one that is very risky and contains uncertain conditions. To address that challenge, planners look more for technological progress that would be able to deal at least with some more pressing issues; to focus on candidate's selection and training; and even human enhancement systems that could render astronauts more resistant and fitting to those harsh conditions. In any case, those conditions will require a variety of coping strategies helping to balance the negative effects associated with that very contingent or unpredictable setting.

13.2 Does Religion Have Still a Function in New and Future Scenarios?

The role religion might play in very secularized societies and in future well-informed and functioning contexts is a matter of contention, both from an empirical and a theoretical point of view. The Martian mission and station become an extreme case to test to what extent religion might still survive and play an effective role in very special settings. At least four different levels deserve to be explored when dealing with that question: the social–theoretical treatment regarding religion's function in advanced societies and contexts; the issue about religious coping and its effectiveness in contemporary societies; religions ability to enhance prosocial behaviour; and beliefs and believing as unavoidable and useful cognitive functions.

13.2.1 Religion's Function in Advanced Societies

For the first level, we cannot go into the current and very complex discussion on secularization or religious crisis in advanced societies. The main or driving idea is that such a loss of religion is due to modernization, progress and people's ability to cope with their own problems and wishes using rational and technical means. Religion is no longer needed and is perceived as less effective or practical when dealing with social and personal issues that in the past could justify religion's existence and utility. However, that may provide some rationale for a possible religious provision or dimension in new planetary settings. A possible approach to the question is to assess which functions are still interesting or relevant when trying to justify religion's presence and recourse and also where it becomes no longer a necessary part of the ideological and imaginary landscape in advanced societies.

Functionalist studies of religion have applied a methodology that looks at religions' utility and performance in society and for individuals, as a way to explain its persistence and working. The American sociologist Talcott Parsons offered useful indications in this regard. For him, 'religion helps individuals to cope with unforeseen events and uncontrollable outcomes' (Parsons 1967). Linked with that function, religion provides 'the primary source of meaning' and moral values for individuals. At a social level, religion is an integrating factor and provides legitimation for the social order (Parsons 1951).

The German sociologist Niklas Luhmann has followed quite closely Parsons footsteps applying a functionalist system theory to analyse religion and its function in modern societies. He offers a well-known description: religion's function consists of determining what is indeterminate or contingent, and to reduce excessive and unmanageable complexity in social processes (Luhmann 1977). Others have commented on these ideas with other words: religion works 'absorbing the risk of failure inherent in all social representations and determinations' (Beyer 2000, 99). Luhmann has tried to deepen such ideas pointing to religion's ability to 'de-paradoxize' or to hide paradoxes arising in other systems, born from their self-referential character (Luhmann 1985). In any case, its 'performance' appears as a service that helps other social systems to cope with their own residual or hard to manage uncertainty, failure and what is excluded or discarded as a consequence of their own selection processes.

The question now is to what extent are the described functions still useful and even necessary or convenient, when societies possibly have learned to deal with their own problems and uncertainties in a more secular or immanent way, without resorting to 'external' sources of meaning. Well, the question is quite easy to answer and can find only a practical or empirical treatment: as far as societies manage to organize themselves in an efficient way, and individuals achieve autonomous or immanent ways to cope with distress and other problems satisfactorily, then religion and its function becomes redundant or obsolete, being replaced by more effective or less costly means—in cognitive terms. However, every observer can see the difficulty in applying such criteria, since much depends on whether we can compare and empirically assess such effectivity and well-functioning in advanced social systems, their smooth integration, and check for involved factors and consequences. In some way, we can imagine that such a state in which uncertainty and failure will be completely determined would render religion superfluous. However, this appears very unlikely, and this perception is increased when humans undertake more risky endeavours where contingency levels increase, and outcomes are more uncertain.

Sociologists studying religion have observed in the last decades that its impact and significance is inversely related to levels of so-called 'existential security' (Norris and Ingelhart 2004). This claim can be understood as a counter-probe for the former functionalist descriptions. If religion helps to cope with uncertainty and unresolved troubles, then a social system in which such risky factors decrease can expect a religious decline. Now the inverse tendency needs to be assessed: whether religion's demand would grow when indeterminate and intractable circumstances are threatening personal life and social stability. The point now is that such fluctuations can be assumed and appear as quite consequential; but since there is no way to eliminate entirely future uncertainty or failures, then we cannot imagine a scenario in which we could get rid of any religious rest. Our fear could worsen in the current environmental and political conditions, eliciting very negative outcomes for everybody.

13.2.2 Religious Coping

Moving to a different plane, religious coping becomes relevant in our research as a way to access religious positive functions for individuals and their well-being. Indeed, the psychological problems raised in the former paragraph point to coping strategies that might be able to deal with the expected anxiety, boredom, stress, loneliness and threats that would probably afflict even the strongest and best trained personalities involved in interplanetary expeditions.

After the studies promoted by Pargament and his team (Pargament 1997), countless empirical and experimental research has been conducted and published on 'religious coping' and how that works when dealing with many physical and psychological sources of distress and dysfunctional behaviour. Now some issues are relevant for our own attempt at considering religion as a trusted resource, mostly ignored in the current literature on spatial missions, but widely recognized as a valid and positive strategy. The first issue regards the effectiveness of such strategy and the conditions for it; and the second asks whether it can still make sense or might it be easily replaced by psychotherapeutic means.

Broadly speaking, most studies have shown a moderate positive effect of religious beliefs and practices on psychological adjustment and dealing with stress and depression (Ano and Vasconcelles 2005). However, some caveat needs to be mentioned: that positive effect depends on positive religious forms, and is reversed when religion assumes a negative, punishing character. Probably more depends not only upon those very general religious orientations, but on religious intensity and commitment levels and related factors, like community support and practices or styles. This is not the place to deal with the immense literature and its many nuances. The point is that some religious forms or expressions show a helpful effect when dealing with negative life events or periods, but the evidence is neither massive nor conclusive; the effect is just moderate and depending on several factors. Then, the question is raised about its convenience, where a substitution for religion might be more practical. Many people resort to alternative or secular coping strategies when they confront stress or depression. In a recent study on adolescent coping strategies, youngsters resorted to both religious and secular strategies and combined them without problems or tensions (Torralba 2018).

It is reassuring to know that religious coping strategies are still available, even in secularized societies and that they are quite effective when dealing with psychological distress and related problems, even if they do not replace or displace more focused or professional therapies but are seen rather as complementary or supporting strategies. In that sense, to ignore religious therapeutic effects in advanced societies would be a bad idea, according to the American anthropologist Luhrmann (Luhrmann 2013).

13.2.3 Religion and Prosocial Behaviour

The third point to consider is to what extent religion becomes a resource in modern societies or even future settings because of its prosocial-enhancing capacity. This issue has drawn considerable attention and more than one hundred studies in recent years—empirical, experimental and theoretical—trying to better understand to what extent religion is still a factor able to enhance prosocial behaviour. Prosocial behaviour would be highly desirable for astronauts. The outcomes of these studies are mixed and not easy to apply, as is evident from available systematic reviews and meta-analyses (Galen 2012; Sablosky 2014). The main factor weighing on prosocial behaviour seems to be altruistic personality, as can be expected. Religion still plays a role, but once more some caveat is required. Not every religion promotes interest in helping others, or an altruistic inclination, but only religions more oriented towards moral commitment for neighbours or more compassion, like the 'Post-Axial' religions. Then, the ability of religion to bring out altruistic behaviour depends on other factors, and—very relevant—often that effect works only towards in-group members, and less towards strangers or those belonging to other religions, when it can turn to hostility.

The available evidence shows that some religions promote prosocial attitudes and help to build stronger communities and bonding links among members, rendering them more resistant and adaptive. But, then, altruism works in some cases as a boundary enhancer against other social groups seen as rivals. How effective and enduring is the effect depends on many circumstances, and often it requires a strong religious commitment. Religion, devotion or spirituality per se do not appear to entail an interest in helping other people, but in more cases leans towards an inward and self-serving attitude. Then, recent altruism studies have shown that such prosocial behaviours can be associated with empathy and education and are less or least linked to religious ideas and practices per se (Martin 2016).

13.2.4 Beliefs and Believing

A step further invites us to consider beliefs and their functions so as to better explain the religious mind. Recent research has firmly established that beliefs are cognitive contents that help to better represent our world and to orient our action, assisting in making better decisions (Seitz and Angel 2014; Connors and Halligan 2015; Smith 2016). Beliefs come in degrees and assume a variety of expressions. Against a former tendency that has disqualified beliefs as a source of pathologies and weird ideas, this new wave focuses on their positive effects and convenience, avoiding a competition with other cognitive expressions that provide greater certainty. Indeed, all the discussion around bias and heuristics teaches about the difficulty in reaching such certainty levels and the need to resort to uncertain or tentative views, values and assumed schemas that help to deal with day-to-day and long-term decisions.

Once the positive role of beliefs is assumed, the question still pending is to what extent would religious beliefs be welcomed in a highly informed and critical culture, where some beliefs are more difficult to acquire and hold than others. The discussion in this case is also contextual and practical. Studies of beliefs help to reframe the issue about religion's relevance in advanced settings. In short, this research points out the necessary role played by some beliefs that can be designed as 'ultimate' and that provide meaning and purpose, as unavoidable resources to conduct human life. The recent attention being paid to purpose and meaning provision in psychology and social sciences is a revealing sign of the importance of a somehow neglected topic in contemporary research (Hicks and Routledge 2013; Froese 2016). In any case, after stressing the cognitive belief function, and the importance of a subclass, whose role is meaning and purpose provision, it is still doubtful that religious beliefs will become the most fitting factor to play that central role and to assist in life's direction, nourishing hope and sustaining values and moral commitment. The open question is still whether secular and humanistic versions could work better than religious beliefs after dropping transcendence or reference to external or alternative realms of reality. The question can be posed in different terms when it is accepted that ultimate values and beliefs are unavoidable, and that ultimacy or absolute character confers on them a tone of transcendence, or an almost religious appearance. Maybe we can avoid the issue concerning religious dimension in a Mars mission and station, but hardly will we be able to avoid the issue about the ultimate values that justify such a choice, the possible sacrifice of a life put at great risk or breaking probably in a complete way all ties with all relatives and friends.

Summarizing the points raised in this subsection, the discussion about religion's function and meaning in advanced societies is far from being settled. At an empirical level, available data reveal a religious decline in most Western countries, or those more developed, secure and protective societies. A different issue is to what extent the mentioned discussion and the dismissive data can be transferred to a quite different context, where some humans will have to cope with new conditions and risks, with elevated uncertainty levels and isolation. The prospects point towards a selected crew from very secularized social backgrounds, highly scientifically cultured and closer to social segments quite removed from religious beliefs and practices. However, the described issues should help to connect the religious supply and function in advanced social settings with the expected needs we can expect will exist in that hypothetical scenario, highlighting functions and performances that could turn out to be useful when undertaking such an endeavour. In this scenario of a Mars mission, a highly technological and mission success orientation will still have to be connected with extremely risky and unpredictable environment, a completely new and untested situation which could render religious means more meaningful than in the respective cultural backgrounds of the astronauts, even those who on earth were more secularized.

13.3 Which Religious Format Would Be More Fitting for a Mars Colony?

Religion appears in many forms or kinds, in the historical record and in the present expressions. Typologies can be found out or designed depending on chosen criteria. The supply is vast and somebody looking for the most fitting religious form in a Mars settlement would suffer embarrassment due to such a large number of current religions from which to choose.

After a consideration of all of the probable psychological and social problems that Mars inhabitants may have to endure, an engineering operation would try to design the best religion for those conditions, after assuming that believing some creed and practicing some rituals would be after all a good idea in order to cope with foreseeable distress. This is probably a modern predicament. After many centuries, it seems that we can choose and even build the most convenient religion depending upon the context and practical needs, or perhaps upon culture and mentality. There is now an undeniable pluralism in earthly religions. Such a pluralism renders even Christian faith quite flexible and adapted to every social and cultural need. There are also earthly religions with a culture of conservation or one of protest and social changes, to traditional views and to more liberal ones. And then there are earthly religions with emphasis on feminism or race or sexual orientation. There are religions that are more adapted to African, Asian or Latino cultures. Why should a new planetary context become an exception? Adaptability—instead of rigidity—seems required for every religious expression that looks for survival. Nevertheless, some questions still arise regarding that tendency of adaptability, as will be exposed in the next paragraph.

At the moment, what we can do is to look for religious forms that could be more attuned with the type of life that those living on Mars will have. The eremitic life, living a simple life, sometimes as a solitary hermit, has very ancient roots and is shared by several religious traditions, not only Christianity, but this type of lifestyle is present in Hinduism, Buddhism, Islam and still more in Daoism. In all these cases, individual persons inspired by their religious beliefs leave their own society to live isolated and on their own. In many cases, those solitary persons were congregating and forming communities. In other cases, hermits were gathering in a group or forming a monastic community. These earthly eremitic approaches to religious experience could be inspiring for a remote settlement on Mars, where eremitic life would be more pressing and demanding, and even more exposed to a mysterious Providence. However, probably here we finish the analogy and simile, since the reasons and motivations appear quite different and even opposed. Then, since eremitic life is shared by several religious traditions, we would still have to choose the most convenient one for a Mars settlement.

A way to address this very practical problem could be to start with some distinctions that could help to better build a fitting pattern. Observation and traditional sociology of religion can be useful. For instance, there is a constant distinction between the active religions and the more mystically driven, or passive ones. As a first choice, one must try between a religious inspiration that stresses more the practical engagement with the world, as Max Weber described (1920); or rather a passive and spiritual practice aimed at meditation and quietness. This is a characteristic in several evolved religions, where such double path is present and offers different engagement forms. To this distinction Max Weber added a third type, that appears as a subclass of the active dimension: the religion of fraternity, more committed to improve brotherhood and social justice. We can envisage even some distribution of religious roles in that imaginary colony: some adopting an active and others a mystical form, since both make sense in that extremely difficult context. Other distinctions offer a spectrum between a greater or lower intensity of commitment. This is not a trivial question, since in several cases, religion's functionality works only when a sufficient intensity is reached.

Obviously, the existing religious traditions on earth present some possibilities for a religion for Mars. Trying to imagine some axis for the spectrum of religions on earth, we could assume an elementary distinction between two paradigmatic religious forms: Buddhism and Christianity. Even if a combination of these two religions appeared feasible, it is more realistic to assume that one or other by itself would make more sense, considering the different challenges and conditions of combining them. Perhaps, which earthly religion one adopts is a personal choice and cannot be addressed from a more 'technical' point of view as to which religion would be more fitting for the Mars environment.

At this point in time, it seems wiser to offer the future Mars inhabitants information and a spectrum of choices from which they can choose, depending on personality, cultural background, values and preferences. In any case, the religious style—besides the nominal tradition or confession—should be such that it might assist in coping with high contingencies, psychological stressors and loneliness. A Martian religion, whatever it might be, should help to build strong social bonds and to enhance prosocial attitudes.

A related aspect we cannot neglect is the communal aspect of religious expression. If one's religion is left to personal choice, it would be very difficult to organize a minimal of astronauts for shared rituals or prayers and religious services. A kind of communal spirit informed by a deep ecumenical commitment would impose not only respect for any other religious traditions, but even the participation in others' rites and sharing or supporting their practices. Religion which would not be just private if the astronauts have to work together as expected. On a practical level, such a context might invite religion to become at the same time personal or intimate, and also shared and expressed openly with other crew members. Those conditions require a respectful attitude and open dialogue, and would have to exclude religious forms that become more closed, exclusive and fanatic. In this case, selection and training becomes important. However, those designing the Mars program or architecture need to be aware of the possibility that religion could contribute to the mission's success, considering its functions, but—at the same time—the problems it could entail. It would be not a clever idea—in order to avoid conflicts—to select astronauts who are agnostics or those less religious. A better solution would point to those who are religious and open or able to assume and integrate other religious or spiritual forms with their own. Selecting such astronauts should not be such a difficult task. There should be plenty of religious astronauts in contemporary societies and cultures on earth, provided that we assume that there is a positive function for selected religious expressions on the Mars mission.

The suggested practice could posit a dilemma between the free choice of religion and some obliged and forced choices among potential astronauts during the selection process. In a possible long-term scenario, when Mars will know a permanent human settlement, people will probably reproduce and give birth to offspring there (Szocik and Marques 2018). Then the question will arise about how and what to teach those children to live on Mars, to cope with stress and anxiety, and to behave prosocially. Again, the issue is to what extent religion can play a positive role, and which kind of religion could be specially designed for people born on Mars and fitting in that environment.

13.4 Looking for the Right 'Spatial' Religion: Does It Make Sense After All?

Religions have been created or invented at a high rate during history on earth and still more are being created in the current time. Creativity is a central characteristic of human religious genius, and so it should not be a surprise that we envisage religious forms tailored to a long-term space station on a remote planet. It is possible that a religiously inspired person proposes a new and original expression of an earthly religion; and another proposes to engineer from scratch a religion designed to fit in the peculiar conditions of a Mars colony.

Some issues arise when trying to conceive or create an ad hoc religion, or one just adapted to new circumstances. Since religion determines ultimate meaning and values, it appears as an oxymoron that somebody can design his or her own religion. It would be a self-serving approach, and certainly it would not draw followers or draw any attention.

August Comte was already trying to build in the nineteenth century a 'positivist religion' tailored to a time in which science assumed a leading role and became a source of value. This attempt was placed in continuity with former trials during the French Revolution to build a religion that rendered cult to reason, or a 'rational religion'. That programme found some continuity in later efforts to design a religion linked or built on nationalism, on science, on art or emotions. Such programmes have always failed, and for a similar reason. We cannot freely dispose what per definition is unavailable; we cannot create whatever that by its own nature is uncreated; or cannot serve ourselves of what is beyond any will. In that sense, religion can hardly be the product of a tailoring or designing activity, instrumental to one's own needs. In traditional theology, these attempts have always been denounced as 'idolatry', or efforts at building a God that serves a population and their needs, instead of looking for religious expressions that could really 'save' and help to overcome people's deep needs and overcome their worst evils. What is called for in contrast is a 'true religion' that reflects rather a spiritual search that looks for truth and illumination.

On a practical level, a project that contemplates religion in a Martian crew needs to start with preparing astronauts for such searching activity. They need to be provided with first quality religious resources or inspiring texts and that will orient them on that spiritual journey. A hypothetic scenario would be that humans going through such an experience could feel special inspiration or be illuminated in such extreme circumstances to be able to conceive a religious understanding that could be more fitting for that very special context. The selected crew could count on one or more individuals attuned with that sensitivity and able to develop spiritual answers in completely new contexts.

In general, religious genius has worked on the shoulders of former available traditions which have been re-elaborated or extended in connection with new circumstances and experiences. As a consequence, we are in no position to build a religion for astronaut's consumption, some set of beliefs and practices carefully packed and well-integrated. What we can provide are some general orientations that allow other-planet settlers to undertake a deep spiritual journey aimed at finding out the more fitting religious expressions in that new setting, expecting that inspiration, trials, failures and some positive outcomes or experiences could help to delineate—in a kind of spiritual dialogue with transcendent dimensions—their religious path. After all, this has been the shared characteristic of many religious seekers and most new religious founders or creative spiritual minds. They have lived their experience as a path and a dialogical process with the Divine, which was taking place through events, internal and external experiences, deep intuitions and setbacks that invited them to correct their route and to explore alternative territories and opportunities. Since Mars is still *terra incognita*, it becomes hard and even reckless to try to anticipate and foresee how religious feelings, thoughts and emotions could be deployed and developed in that context. It will be wiser to leave open the possibility for exploration and discovery. Indeed, a Mars mission could become, from this point of view, an opportunity not just to explore far and strange worlds, but to explore our own spiritual capacities in a very different context, and how humans can grow and discover new spiritual dimensions when our conditions change in a radical way, as has happened many times in the past.

13.5 Concluding Remarks: Adapting Religious Minds

So far, an analysis of the conditions that could justify looking for religious assistance in a future expected Mars station has been presented. Being aware of the stressors, uncertainty, motivational and ethical needs associated with that programme, religion can be described as functional and fitting to address these challenges and provide effective coping strategies. Describing the utility of religion in such unexplored contexts still raises questions about the most convenient religious form, beliefs and practices that could become instrumental for such mission. Apparently, not every religious expression on earth today would satisfy the described requirements. The multiple choices available should not discourage, but rather stimulate to look for models that support spiritual search and exploration as the most fitting religious style for the Mars mission, instead of closed patterns based on existing rigid traditions. Since religion—as any other cultural expression—evolves and adapts following its own dynamics, so the new Martian context will provide a new environment in which religion could evolve and adapt from existing available models, but following again a logic similar to other processes of cultural evolution and adaptation (Richerson and Boyd 2005; Richerson and Christiansen 2013; Laland 2017). Past experiences about such evolutionary processes could guide the new spiritual exploration, but more than that, the question remains necessarily open and inspired by hope.

References

Ano, G. G., & Vasconcelles, E. B. (2005). Religious coping and psychological adjustment to stress: A meta-analysis. *Clinical Psychology, 61*(4), 461–480.

Beyer, P. (2000). Religion as communication: On Niklas Luhmann, the religion of society. In M Stausberg (Ed.), *Contemporary theories of religion: A critical companion* (99–115). London: Routledge.

Connors, M. H., & Halligan, P. W. (2015). A cognitive account of belief: a tentative roadmap. *Frontiers in Psychology, 5,* Article 1588. https://doi.org/10.3389/fpsyg.2014.01588.

Do, S., Owens, A., Ho, K., Schreiner, S., & de Weck, O. (2016). An independent assessment of the technical feasibility of the Mars One mission plan—Updated analysis. *Acta Astronautica, 120,* 192–228.

Froese, P. (2016). *On purpose: How we create the meaning of life.* Oxford: Oxford University Press.

Galen, L. W. (2012). Does religious belief promote prosociality? A critical examination. *Psychological Bulletin, 138,* 876–906.

Goemaere, S., Vansteenkiste, M., & Van Petegem, S. (2016). Gaining deeper insight into the psychological challenges of human space flight: The role of motivational dynamics. *Acta Astronautica, 121*(2016), 130–143.

Hicks, J. A., & Routledge, C. (Eds.). (2013). *The experience of meaning in life: Classical perspectives, emerging themes, and controversies.* Dordrecht: Springer.

Kanas, N., Sandal, G., Boyd, J. E., Gushin, V. I., Manzey, D., Northf, R., et al. (2009). Psychology and culture during long-duration space missions. *Acta Astronautica, 64,* 659–677.

Kanas, N. (2011). From Earth's orbit to the outer planets and beyond: Psychological issues in space. *Acta Astronautica, 68,* 576–581.

Kanas, N. (2014). *Psychosocial issues during an expedition to mars, in the new Martians: A scientific novel*. Heidelberg, Dordrecht: Springer. https://doi.org/10.1007/978-3-319-00975-9_2.
Laland, K. (2017). *Darwin's unfinished symphony: How culture made the human mind*. Princeton, NJ: Princeton University Press.
Luhmann, N. (1977) *Function der religion*. Frankfurt a.M.: Suhrkamp.
Luhmann, N. (1985). Society, meaning, religion- based on self-reference. *Sociological Analysis, 46*, 5–20.
Luhrmann, T. M. (2013). When god is your therapist, *New York Times*, April 13, opened in: http://www.nytimes.com/2013/04/14/opinion/sunday/luhrmann-when-god-is-your-therapist.html.
Manzey, D. (2004). Human missions to Mars: New psychological challenges and research issues. *Acta Astronautica, 55*, 781–790.
Martin, J. (2016). Advances in altruism studies: Their relevance for theology. *ESSSAT News & Reviews, 26–1*, 5–18.
Nair, G., Madhavan, K. R. Sridhara, & Murthi, M. Y. S. Prasad. (2008). Strategic, technological and ethical aspects of establishing colonies on Moon and Mars. *Acta Astronautica, 63*, 1337–1342.
Norris, P., & Ingelhart, R. (2004). *Sacred and secular: Religion and politics worldwide*. Cambridge: Cambridge University Press.
Pargament, K. I. (1997). *The psychology of religion and coping: Theory, research, practice*. Guilford Press: New York, London.
Parsons, T. (1951). *The social system*. Glencoe, Ill: Free Press.
Parsons, T. (1967). *Sociological theory and modern society*. Glencoe, Ill: Free Press.
Richerson, P. J., & Christiansen, M. H. (Eds.). (2013). *Cultural evolution: Society, technology, language, and religion*. Cambridge, MA, London UK: MIT Press.
Richerson, P. J., & Boyd, R. (2005). *Not by genes alone*. Chicago: University of Chicago Press.
Sablosky, R. (2014). Does religion foster generosity? *The Social Science Journal, 51–4*, 545–555.
Sandal, G. M., & Bye, H. H. (2015). Value diversity and crew relationships during a simulated space flight to Mars, Acta Astronautica, *114*, 164–173.
Seitz, R. J., & Angel, H.-F. (2014). Psychology of religion and spirituality: Meaning-making and processes of believing. *Religion, Brain & Behavior, 5–2*, 139–147.
Slobodian, R. E. (2015). Selling space colonization and immortality: A psychosocial, anthropological critique of the rush to colonize Mars. *Acta Astronautica, 113*, 89–104.
Smith, A. C. T. (2016). *Cognitive mechanisms of belief change*. Basingstoke, UK, New York: Palgrave Macmillan.
Šolcová, I. P., Šolcová, I., Stuchlíková, I., & Mazehóová, Y. (2016). The story of 520 days on a simulated flight to Mars. *Acta Astronautica, 126*, 178–189.
Suedfeld, P. (2005). Invulnerability, coping, Salutogenesis, integration: Four Phases of space psychology. *Aviation, Space, and Environmental Medicine*, 76(Suppl. 1), B61–B66(6).
Suedfeld, P. (2006). Space memoirs: Value hierarchies before and after missions—A pilot study. *Acta Astronautica, 58–11*, 583–586.
Suedfeld, P., Brcic, J., & Legkaia, K. (2009). Coping with the problems of space flight: Reports from astronauts and cosmonauts. *Acta Astronautica, 65*, 312–324.
Szocik, K. (2017). Religion in a future mars colony? *Spaceflight, 59*, 92–97.
Szocik, K., Marques, R. E., et al. (2018). Biological and social challenges of human reproduction in a long-term Mars base. *Futures, 100*, 56–62.
Tachibana, K., Tachibana, S., & Inoue, N. (2017). From outer space to Earth—The social significance of isolated and confined environment research in human space exploration. *Acta Astronautica, 140*, 273–283.
Tafforin, C., & Abati, F. G. (2017). Cultural ethology as a new approach of interplanetary crew's behaviour. *Acta Astronautica, 139*, 102–110.
Torralba, J. (2018), Religious coping in adolescents: New evidence and some nuances, *under submission*.
Weber, M. (1920). Zwischenbetrachtung, *Gesammelte Ausätze zur Religionssoziologie I*, Tübingen: Mohr, 536–573.

Chapter 14
Human Place in the Outer Space: Skeptical Remarks

Konrad Szocik

Abstract The most skeptical contribution to this volume enumerates and discusses a broad set of challenges connected with the so-called human factor in a mission to Mars. Discussed issues include rationales for a human versus uncrewed mission (the chapter suggests that human missions could be successfully replaced by robotic missions), financial challenges affected mostly by unclear and weak rationales for human mission, challenges of sustainable development, complex hazardous impacts of space environment for human mental, and physiological health. The last of the discussed challenges, the idea of human enhancement applied for the purpose of human deep-space missions, shows how technological issues—mostly long journey or ineffective countermeasures—might affect ethical concerns. While this idea might seem to be too far in the future, the chapter shows that it may be a serious and possibly unexpected long-term consequence of this program. This chapter does not determine whether a human mission to Mars is possible or not, nor whether such a mission makes any sense at all. One side of this chapter assumes that it is hard to find a strong rationale as measured in financial terms. The question of rationale is getting harder when a cost–benefit analysis—including risks for human health and life—is applied. On the other side, these skeptical remarks are designed to show that the idea of making humans a multi-planetary species is, in fact, extrapolation and projection of all problems and challenges known on Earth, which is intensified by putting Mars astronauts in the hazardous space environment.

14.1 Introduction

Should humans go to Mars or not? Is there possibly one conclusive answer? Advocates and opponents of human interplanetary missions are trying to offer more or less decisive answers. Advocates of human mission to Mars seem to believe in the power of science, technology, human mind, and progress. Opponents are a more diverse

K. Szocik (✉)
Department of Philosophy and Cognitive Science, University of Information Technology and Management in Rzeszow, Rzeszów, Poland
e-mail: kszocik@wsiz.rzeszow.pl

K. Szocik (ed.), *The Human Factor in a Mission to Mars*,
Space and Society, https://doi.org/10.1007/978-3-030-02059-0_14

group. Some of them treat the idea of a human interplanetary mission in terms of science-fiction fantasy. Others are not able to imagine any rationale for going there even if the state of the art in space technology would make possible such a mission. This a is trivial, but important distinction. Possessing technology that makes possible interplanetary missions does not make them reasonable and necessary. This is the case for human missions to the Moon. NASA's human lunar program *Apollo* was realized only in 1966–1972. When the political aim of the US competing with the Soviet Union was achieved, from 1972 until today, no strong rationale appeared which would be able to give the incentive to send humans once again to the Moon. All rationales, including benefits for science or an opportunity to test and to develop space technologies, may seem to be weak when compared with existential risks on Earth, for instance. Other subgroups of critiques of human interplanetary missions do not find any benefits. Cost–benefit analysis of human mission to Mars is their main counterargument. Cost–benefit analysis includes not only financial costs and financial profits, but all other possible benefits and costs.

This chapter discusses the possible benefits and costs expected to be a part of a human mission to Mars. Possible rationales for and objections to this mission are discussed, mostly in the context of the human factor. In conclusion, this overview suggests that the combined impact of various factors connected with the lack of strong rationale for this mission may challenge its realization.

14.2 Human Space Missions—The Brief Remark

This very brief historical remark shows how a big challenge for humanity must be launching a human interplanetary mission. No space agency has made progress in human spaceflights from the times of the first human landing on the Moon (Apollo 11 in 1969) and the first space station (Salyut 1 in 1971). The current human space program includes the only service at the International Space Station (ISS). No permanent or semipermanent human base has been ever built on the Moon. It is worth mentioning the fact of cancelation in 2011 of the American program of space shuttles operated on Earth orbit. In terms of distance from Earth, human achievements are poor. No progress has been made since the 60s and 70s. Human astronauts do not explore or discover new places in space. They travel in orbit around Earth and do not go beyond.

This remark on the current state of the art, mostly when compared with the past crewed lunar missions, is not intended to criticize space agencies or humanity in general for their limited capabilities in the exploration of space. But this historical remark implies two thoughts. One of them assumes that the lack of progress in human space missions is caused by technological limitations. The basic one includes the need for survival of the crew during about 3 years mission to Mars. This is the shortest variant of the mission to Mars, which does not consider a longer stay in a research base, or permanent or semi-permanent settlement. It may be expected that any longer variant will cause more and more technical challenges connected with efficiency of the life support system, supplies from Earth or in situ resources utilization, and protection of the crew against the hazardous space environment.

The second thought assumes that technological infeasibility is a secondary factor. The primary factor is the lack of strong rationales for a human mission to Mars. This is only a hypothesis, which should be considered and discussed in depth. Human missions to the Moon did not have any rationale excluding political value of the space race. They brought scientific results, but progress in science was a by-product, not the target of *Apollo* program. Because the current competition between spacefaring countries including the US, China and Russia—other space actors like Europe, Japan, India and Canada have lower potential and lower capabilities—refers to terrestrial policy, there are no political incentives to develop space programs like during the Cold War.[1] It is worth keeping in mind that an important incentive for space programs in the 50s and 60s was connected with the novelty of a human presence in space. Public opinion—in contrast to the first decades of the twenty-first century—was interested in pioneering space achievements.

Today, when the global political situation offers arguments against, rather than for, space exploration—because of overpopulation and climate change but also increasing ecological awareness, which may inhibit such putatively extravagant expenditure like the human interplanetary program—human astronauts do not explore the interplanetary space, because the rationale to do that is not clear. When a clear and strong rationale is not available, there are no incentives for substantial progress in space technologies enabling an interplanetary journey and a space habitat on another planet.

Technological feasibility is only a partial objection to progress in space exploration. The decrease in human space missions is connected with growing success in robotic space missions. These two branches of space exploration are connected. Funds invested in the human space program limits progress in robotic missions, and vice versa. Robotic missions are a prelude to future human missions. Progress in uncrewed missions is incomparable with the state of the art in the human missions. One of the examples is the mission of the robotic spacecraft Hayabusa operated by JAXA, which realized the first asteroid sample return mission. Another example is the series of uncrewed missions to Mars within the agenda of NASA Mars Exploration Program. Both kinds of space activities are well beyond the current human spaceflight technologies.

14.3 Human Versus Uncrewed Space Missions

In the current state of the art, robotic spacecraft are the essence of space programs. They provide data important for future human missions to Mars. Discoveries made by space robots may, in more or less direct ways, affect the human settlement of Mars.

[1] American president Donald Trump considers the idea of a space force, and there is also the revival of USSPACECOM. But it seems that these attempts will be limited to Low Earth Orbit.

It is worth keeping in mind two recent discoveries. One of them is confirmation of previous discoveries of the resources of subsurface liquid water on Mars (Orosei et al. 2018). Another discovery has been published by Jakosky and Edwards (2018). They show that the simplest variant of terraforming Mars based on releasing of all resources of CO_2 trapped in Mars will fail. As they estimate, all resources of CO_2 on Mars are only 6.9% of the total amount of gases on Earth responsible for the greenhouse effect. This is well below the minimal level required for producing a warmer atmosphere on Mars. While confirmed resources of liquid water on Mars may support plans of future human settlement, unfeasibility of the easiest technique of terraforming Mars may negatively affect these scenarios of Mars settlement, which assume terraforming of the planet. If this were the case, this fact would definitely slow down the process of terraforming of Mars, which—before discoveries of Jakosky and Edwards—was estimated to be about 100 years (McKay et al. 1991). Like Braddock (2018) and others argue, this discovery may change or, in fact, reject plans of space settlement based on terraforming. As they add, production of conditions that make possible photosynthesis, would require about 100,000 years!

This issue is worthy of a brief digression. While the idea of terraforming is well beyond technological possibilities and a reasonable timeline, the rationale for terraforming of the outer planets is not clear. Warming the atmosphere is not equivalent to the production of an atmosphere containing oxygen. Terraforming will not remove the necessity of life support systems and extra habitat activity that permanently require space suits. It is worth remembering that terraforming does not guarantee safe regolith, which would make plant cultivation, possible. Martian dust and regolith may be toxic, although this fact remains unknown. Philosophy, which lies behind the idea of terraforming, refers to some minimal level of comfort of human life in space. This comfort is, like Margaret Boone Rapaport and Christopher Corbally suggestively show in this volume, something more than psychological deprivations usually registered in space. This is a combined psychological and existential challenge caused by leaving everything that is typical for the human life on Earth (Cohen and Haeuplik-Meusburger 2015). From this point of view, Mars terraforming seems to be justified.

The human space program is a rival for the uncrewed space exploration. Human missions are more expensive and more challenging technologically (however, in a specific sense, robotic mission may be more challenging if they attempt to be like human astronauts!). Preference for only a human space program inhibits exploration of remote locations. Because a human space program involves substantial budget and staff support, a robotic space program could be slowed down. But the opposite scenario is possible as well. It seems reasonable to assume that science policy of a particular space agency based only on human spaceflights—with a highly marginalized robotic space program—would be able to make substantial progress in preparation for human interplanetary missions.

Space policy should be shaped by needs and rationale. The focus of space agencies on robotic mission is the consequence of the fact that the unique currently considered rationale—but even this one is broadly criticized!—is the contribution to space science. The relevance of human and robotic missions to progress in space science is the main criterion for evaluation of upsides and downsides of both kinds of space exploration.

More detailed debate on the rationale for human versus uncrewed space missions is published elsewhere Crawford (2012), Rovetto (2013, 2016), Shelhamer (2017), Szocik (2019a), Szocik and Tachibana (2019), Weinberg (2013). If we assume that the unique rationale space missions are their impact on progress in science, this discussion should compare features and capabilities of both human and robotic astronauts which are relevant for scientific exploration. It is interesting that there are still the same two opposite components at work, which counterbalance higher effectivity of robots over humans, and vice versa. While space robots may be sent to the border of the solar system and beyond—no one must care for the life support system for them—their capabilities are still well beyond human dexterity and intelligence. These limitations of robots are known in space robotics and they appear mostly in such contexts like communication delay or intelligent, dynamic detection and reaction for scientifically relevant objects and phenomena. But current limitations of space robots do not exclude discoveries made by flyby satellite missions, orbiters, landers and rovers.

These weaknesses of both human and robots lead to the following, apparently contradictory, conclusions. On one side, progress in space science will be limited, unless space robots are able to imitate the mentioned human capabilities. Without them, effectivity and performance of space robots always will be below human capabilities. This fact lies behind the concept of the human–robot partnership. Despite the benefits of such joint missions, this approach does not solve the biggest challenge, which is still the presence of human astronauts in the hostile space environment. This is an important weakness of the current state of the art in space robotics, because one of the most important reasons for applying robots in such environments like space or deep seas is the protection of fragile human factor. But insufficient progress in space robotics has led to a situation where one of the considered scenarios of space missions is exploration telepresence—with humans in orbit around Mars directly controlling in real time rovers on Mars surface (Lester et al. 2017). This solution solves problems connected with descent and ascent vehicles but does not protect humans against space radiation and microgravity.

On the other side, an unquestionable advantage of human astronauts over robots in dexterity and intelligence does not imply a rationale for human missions. The state of the art in the space robotics may work as an incentive toward substantial progress in work on dexterity and cognitive capabilities, including artificial intelligence (AI). From this point of view, focus on a human space program would slow down progress in space robotics and, consequently, would inhibit scientific progress. Because—like it is suggested in this chapter and by other authors—the unique rationale for a mission to Mars is its scientific value, the care for progress in science may work as an objection to the human mission to Mars.

It seems that in this comparison, advocates of a human mission to Mars are in a weaker position in argumentation. They seem to be obliged to prove at least one situation in which the presence of human astronauts on Mars would be necessary. At least two such situations might be considered. One of them is the mentioned advantage of human astronauts over robots in the field. Let's assume that space robots fully equipped with human-like dexterity and intelligence are well beyond the current science robotics. This fact opens room for a human presence in space. But in this scenario, new a challenge appears: the justification for progress in space science based on human spaceflight. Weintraub (2018) argues that looking for life on Mars still matters. McKay (2009) points out that the discovery of tracks of life on Mars will have high value for science, mostly if the second genesis of life could be proven. Gonzalo Munévar emphasizes in this volume the scientific value of the study of Mars. If the series of expeditions within NASA's Mars Exploration Program, including the planned mission of Mars 2020 Rover designed for astrobiological research, is not able to definitely confirm or disconfirm the tracks of life on Mars, then the discussion on rationales for a human mission should be considered even for these purely scientific purposes.

But two counterarguments appear here. First, Mars seems to be a target that is relatively easily available for robotic exploration. The mentioned two examples—available resources of CO_2 and the subsurface ocean of liquid water—or the series of experiments designed to detect methane (Weintraub 2018) show that progress in astrobiology research on Mars is possible without human astronauts. While the advantage of human dexterity and intelligence discussed above is relevant also to Mars, these capabilities are more important in robotic exploration of more remote locations that are not available for landers and rovers, for instance. But, in any case, human astronauts on Mars would be able to offer the extra value in their dexterity—always connected with cognition used to detect and to find scientifically relevant targets—to collect and to send back Martian samples to Earth.[2] Second, while expected scientific discoveries on Mars may possess high value for the scientific community, space science is not necessarily the most valuable and socially strategic scientific field. If we assume that various anthropogenic risks will grow in the next decades, then investing in the study of Mars does not seem to be relevant for coping with existential risks on Earth.

The next scenario, in which human astronauts cannot be replaced by robots even if technology in space robotics would make it possible, is connected with the evacuation scenario of post-catastrophic Earth. It is assumed here that there are existential risks that would require space refuges. But catastrophic Earth scenarios do not justify space settlement. Space refuges should be considered only as the very final solution. First, all terrestrial strategies should be considered and tested, including anti-risk prevention and terrestrial—subterranean and/or aquatic—refuges. Advocates of a space refuge are obliged to prove that there is at least one kind of existential risk which might destroy all living forms on Earth—or at least humans. As it was discussed in detail elsewhere Szocik (2019b), it is hard to find at least one kind of such existential threat that would appear in a reasonable timeline—excluding the case of the Sun which will get warmer in the next five billion years. It is worth keeping in mind that scholars who study the issue of the future existential risks for humanity do

[2]While space robots with full human-like dexterity and cognition can collect or send back Martian sample on Earth, space robotics is still beyond human-like robots.

not consider space refuges as a remedy. Despite some exceptions like, among others, Smith and Davies (2012) who are advocates of space settlement and argue for human "adaptation to space", it is hard to find in referred academic publications arguments for space refuges.[3] Such considerations appear mostly in nonacademic publications among science-fiction writers, some scientists or entrepreneurs. In recent years, two of the most famous advocates of human species as a multi-planetary species are Stephen Hawking and Elon Musk, who explore the post-catastrophic scenario. But, as it was mentioned, permanent living in space refuges, dependent on life support systems, might in fact protect life, but it would reduce human life to a simple biological existence. If a terrestrial refuge is able to protect against possible anthropogenic and exogenous existential risks—it is assumed here a scenario in which some part of the human population always survives every catastrophe—then living in a refuge on Earth is always better for a species adapted to earthly conditions. Even post-catastrophic Earth offers 1 g gravity and anti-radiation protection—basic factors unavoidable on Mars. A space refuge is the final option. However, this would not have to imply necessarily that the option has less rationale if we agree not only that Earth will end its life in the future but also that the space-scientific advancements in space refuge technology will provide various high-quality technologies that will allow us to survive in severe terrestrial environments.

This second rationale for human presence in space is doubtful not only due to the fact that probably no kind of risk would require to search for refuges and shelter on Mars. The permanent psychological and existential deprivation which that be a domain of living in Martian shelters, due to the mentioned challenges of the process of terraforming Mars, will be probably a permanent way of living on the outer planet. It is worth keeping in mind that the process of production of Earth-like oxygen-rich atmosphere is estimated to require about 100,000 years (McKay et al. 1991).

As it was shown, it is hard to offer strong reasons for human interplanetary missions. Besides human spaceflight, there is a strong robotic alternative for a human presence in space. Possible reasons for human settlement on Mars are also minimized by factors like financial limitations of public budgets and directions in public space policies, sustainable development, future social and political stability, or the hazardous impact of space environment on human health and life.

14.4 Financial Costs of Space Program

The human place in outer space should be discussed also in financial terms, because the main sponsor of space programs is the public budget of space agencies. Private investors are not considered here for a couple of reasons. First, their budget is well below the minimal safe level needed for pre-launch research and development,

[3]Some brief mentions about space refuges may be found in academic papers, but usually at the margin of other considerations. In contrast to them, it is hard to find scientific publications, which argue for space refuges based on cost to benefit analysis.

launching the mission, and maintenance of the mission. It is worth mentioning that failure or catastrophe may slow down or even cancel private space projects, while public space agencies are not financially limited in this sense. Second, a human space program—at least in regard to interplanetary missions—does not offer commercial applications, which would be able to justify an expensive investment. Private investors by definition require profits, at least in the capitalistic, free-market sense of this term. This remark relates to the third counterargument: the space program is a long-term effort and is not compatible with short-term policies of private investors. Fourth, the level of risk and unpredictability is too high for private investors. While private investors may—and they really do—invest in low Earth orbit activities or try to go beyond, like in the case of planned space tourism around the Moon, private interplanetary plans like, for instance, the Mars One initiative are technologically and financially infeasible (Do et al. 2016). In this case, technological infeasibility is the effect of financial insufficiency. Financial insufficiency works like a vicious circle. Private investors meet the funding wall because their starting budget, which is too small to finance a space program, cannot be supported by the lack of expected benefits. No future benefits are able to provide incentives to invest in these space projects.

But a space program is also a permanent challenge for public space agencies, which—at least to some extent highly extending the capacities of private investors—are unable to get unlimited funding from politicians. A space program—excluding a specific period of the rivalry during the Cold War—is not on the top of the list of the most expensive public expenses. It refers mostly to the human spaceflight program and, first of all, to interplanetary missions. The annual budget of NASA in 2018 is about 0.5% of the total federal budget. For comparison, the annual budget of the US army in 2017 was about 3.1%. The peak of federal support in NASA's history was achieved in 1966, when NASA got 4.41% of the federal budget. The total cost of NASA's InSight Mars lander launched in 2018 is about $814 million. An estimated cost of one USS Gerald R. Ford aircraft carrier is about $13 billion. It means that 16 robotic Mars missions could be launched instead of building one aircraft carrier. Estimated costs of a Mars One one way ticket human mission to Mars is $6 billion. These costs calculated by mission planners are infeasible. Reliable estimates predict at least $500 billion. The annual NASA budget is about $20 billion. Such spacefaring powers as the US or China probably would be able to cover the costs of this project. An international effort would be the best solution but no less problematic as well. One of the basic challenges could be the right of participation for all involved countries in possible benefits of a Mars mission. This topic refers to the broad field of space law, and legal and political challenges mentioned in this volume by Chris Impey. It may be expected that rights of the US or China will be stronger than other countries whose space agencies are able to produce components, but which are not able to produce final products based on expensive and highly advanced technology. In the context of financial costs, two issues seem to be especially worthy of discussion.

One of them is the rationale for human space missions, including a mission to Mars. The annual cost of the International Space Station is about $2.9 billion. While this is a relatively high expenditure—14.5% of the annual budget of NASA (NASA participates in costs of the ISS, but it is not its unique sponsor)—the ISS is currently the unique human space program. On the one side, the ISS is a heritage of the Space Race in the Cold War; on the other side, this is an expression of human technological progress and a symbol of international collaboration. While these historical and symbolical reasons provide a weak rationale, the relatively low costs of this project inherited from the past make it less controversial than any possible future human missions to Mars or to the Moon. But it does not mean that the ISS is not criticized. This project is broadly criticized by scientists representing probably all disciplines of space sciences who question scientific value and benefits of experiments conducted on the ISS. Mostly, they argue that scientific benefits offered by the ISS are controversial in the sense of their novelty and utility (Slakey and Spudis 2008).

The second issue that should be taken into consideration in the context of the rationale for a human mission to Mars, is the set of the expected benefits including spinoffs. While human astronauts have advantages over surface robots in dexterity and collection of geological samples, it is not clear if Martian resources would possess any material value for Earth. Even if there are such resources, it is unclear if the first human expeditions would be able to explore and utilize in situ resources. In any case, the first human missions to Mars are designed for scientific purposes. Because no obvious and clear benefits are expected during a human mission to Mars, cost–benefit analysis may favor robotic missions. It may sound paradoxical that robotic missions are able to generate more spinoffs than human missions. Weinberg (2013, p. 230) explains this apparently nonintuitive regularity and he shows that uncrewed missions produce useful spinoffs in robotics and computer science. Because—as it was reported in the previous section—effective space robots should imitate human capabilities in the field, substantial progress in computer science and robotics is necessary. Advanced robots and AI could be transferred from space industry to other fields. In contrast, possible spinoffs of crewed missions are less useful for humans on Earth, because solutions tested for living in space are designed to work in different physical and environmental conditions. Technological progress in human missions is focused on life support system and human health.

It seems that a cost–benefit analysis of the financial expenditure of a human mission to Mars should favor uncrewed missions due to less costs and more number of indirect benefits expected as spinoffs connected with progress in robotics and AI.

Because space agencies are public agendas, planners of public budgets should follow criteria of social responsibility and sustainability. Like the paradox of technological progress with regard to human space missions assumes (Szocik 2019a), the future expected advancement in space technologies is in inverse proportion to political and social contexts appropriate or neutral for such a mission. It means that probably the best political and social situation to launch a human mission to Mars is today. Growing anthropogenic risks—exogenous risks are not mentioned here, because, in contrast to anthropogenic risks, they are unpredictable and not necessary unavoidable—will require more funds and political activity to cope with them. The human space program in this context—when technology gets readiness in next decades—may lose political and social support and rationale. This phenomenon is discussed from a different perspective by Salotti and Suhir (2014). They note that while the estimated risks in space are really challenging and go beyond the current state of the art in space technologies, the real challenges are not caused by these risks—likely possible to overcome—but by the costs that are required to achieve an appropriate technological level.

Last but not least: in the context of economic objections to a human interplanetary program, it is worth to mention the concept of asteroid mining. While this is not the case of Mars, this activity could be used to provide at least partial funding of human interplanetary missions. But this undertaking is also challenging for a couple of reasons (for more detail, see: Szocik 2019b). Asteroid mining is technologically challenging. One of the problems is the detection of appropriate, valuable objects. Finally, this activity—if launched successfully—may be financially infeasible if it will decrease the market value of currently scarce minerals.

14.5 The Challenge of Sustainable Development

The concept of sustainable development here means a kind of progress and development of humanity that does not threaten the existence and comfort of life nor the entire humanity neither some of its part. This definition is of high value, because development of the West caused poverty and crisis in many other regions of the world. Western countries have exploited populations and resources of their colonies. This was the case of the past colonialist politics of European countries, but this is the case of the current neocolonialist and imperialistic politics of Western financial companies and governments. For these reasons, the current world does not develop in a sustainable way. Interested readers might want to keep in mind some anthropological approaches (Lee 1985; Valentine 2017), which compare future expansion in space to the new epoch in colonialism. Andreas Losch discusses in this volume the issue of the interplanetary sustainability. I refer here to planetary sustainability, which means careful and responsible resources management on Earth, not in the solar system.

Two important terms here are balance and responsibility. Balanced development is focused on human wellbeing and comfort of life, and it should take into account negative environmental and population effects. Responsibility refers to both synchronous and diachronic contexts. Synchronous context means responsibility of developed countries for less developed parts of the world, which are often more or less exploited. Diachronic responsibility refers to the currently living generations and their responsibility to the planet for future generations.

The idea of a human mission to Mars and Mars settlement is connected with the challenge of sustainability. Governments are responsible for the realization of public purposes and for satisfaction of human needs. Public funds should be spent carefully and should meet social challenges. If the total costs of a human mission to Mars—including not only pre-launch research and development, but also many years of maintenance, support, and deliveries—are high (as it was suggested in the previous section, this fact may be one of the main barriers for private investors), some substantial part of the budget of participating spacefaring countries will be obliged to be spent on maintenance of the mission. This challenge is discussed in this volume by Jacob Haqq-Misra. He emphasizes the longevity of the human space program, which in fact will become a multigenerational effort and responsibility. Ideas discussed by Haqq-Misra should be of high value for mission planners and advocates of the human mission to Mars or the human space settlement in general. Haqq-Misra shows that the decision on realization of such kind of program has unavoidable long-term consequences. While he is focused more on financial context and on the need for long-term collaboration, it is worth to extrapolate his considerations to other fields such as responsibility for natural environment, resources, and less developed parts of the world, and priorities in social policy. A human interplanetary program, once activated, will last a couple of decades and will involve time, energy, resources and funds.

It is worth keeping in mind that a permanent or semipermanent base on Mars is a relatively likely scenario. A long-term Mars base will require constant support and utilization of resources from Earth. An associated issue is the unpredictable future governmental politics of spacefaring countries, and unpredictable global political and social situations. Dependent on the particular phase of progress and advancement in a human Mars program, any decision on its cancelation may be more or less difficult and potentially deleterious for people living in a Mars settlement. For this and many other reasons, self-sustainable Mars base/settlement seems to be the most desirable scenario. But even if some space agency decides to realize such project, in its first years—perhaps decades or even hundreds of years—Mars settlement will be based on deliveries from Earth.

One of the particular issues connected with the topic of sustainable development is the number of personnel involved in the project of a human interplanetary mission. Human space programs involve many scientists, engineers, and associated institutions. The *Apollo* program required the employment of 400,000 people, and support of 20,000 firms and universities (NASA). It is hard to claim that the potential and the expertise of workers involved in the space program could be canalized into other, socially more useful and strategical tasks including coping with hazardous effects of the greenhouse effect, for instance. But it is worth keeping in mind this risk when human resources would be disproportionately focused on the space program instead of growing interest in other areas, such as the mentioned existential risks. Not only the free market, but also public policies shape the demand for the particular fields of study for future undergraduates. Two regularities are here at work. First, after success in the early space program, many young people decided to study the fields of science relevant for the space program. Second, without incentives motivating to study and to work in the fields strategic for sustainable development and for environmental protection, fewer people interested in these fields means less useful work done in the strategic fields on Earth.

To conclude, because anthropogenic risks including overpopulation and harmful climate change will increase, they will generate the non-limited financial needs. In this context, any money invested in the space program would be spent on coping with these challenges. But the care for public funds is not the unique threat for the long-term space program. As was discussed above in the context of Haqq-Misra's chapter, a human Mars program will require political and financial stability, and continuity through many years. The current relatively stable political, social, and financial situation, which is promising enough to finance the human interplanetary mission, may collapse when unexpected global events and/or the anthropogenic risks appear. Mission organizers and sponsors might fall in a dilemma between urgent need for coping with the existential needs on Earth, and the duty of the permanent support for a long-term interplanetary space program.

14.6 Human Health and Performance in the Space Environment

The space environment is totally different from the earthly environment. Living there without a life support system is impossible. But even if supported by technology, living in space remains a hard challenge for humans. This naïve truism is getting important when no strong rationale for a human presence in space is offered. The question arises: may we put the human astronauts at the risk without clear motives and benefits? The issue of human health and life in space is considered here not as a challenge for space technology and medicine but as a challenge for space policy and its need for a strong rationale for a human interplanetary program.

It is worth keeping in mind that living in different—let's call them unnatural places, which are not a part of the environment of evolutionary adaptedness—locations is not problematic per se, if humans are prepared in an appropriate way to live there. For instance, living underwater is not a natural human environment, but sailors are able to stay there many months in submarines. Space mission planners are aware of hazards in space environments, but current technology offers only partial protection of the human life. One of the main challenges is the fact that long space flights are qualitatively and quantitatively different from short flights (Beven 2012). A mission to Mars belongs to the category of long space flights. A mission to Mars, when compared with previous long-term missions on the ISS, brings an unpredictable set of risks and threats. Currently, the longest permanent human mission in space lasting 437 days is less than half of the expected period of the shortest, three years human mission to Mars. It is worth keeping in mind that even the longest ISS missions are not affected by some extra risks and loads such as landing and starting from Mars or exposure to still unconfirmed, possibly toxic, Martian regolith. Even if the total sum of time of interplanetary travels will be the same as long permanent missions on the ISS—about 12 months—both missions occur in the different environments. The ISS is protected by earthly atmosphere and magnetosphere and never goes beyond the borders of the low Earth orbit. Interplanetary space is a more dangerous place because of high exposure to space radiation. The next hazardous factor is microgravity, which may strengthen deleterious effects of space radiation. However, there are contradictory data about the deleterious correlated impact of microgravity and space radiation (Ohnishi et al. 2002, p. S9).

The risk to human health and life is an obstacle for a human mission to Mars. If there are no other rationales than progress in science, then sending humans to Mars might be perceived as a risky and irresponsible extravagance. Two kinds of space hazards seem to be exceptionally challenging and still unavoidable, namely space radiation and psychological problems. The third space hazard, altered gravity, is able to cause serious effects as well, but can be, in a relatively effective way, countermeasured by physical exercises.

14.6.1 Space Radiation

The highest exposure to space radiation occurs during an interplanetary journey, because humans leave the protective shielding of Earth. Space radiation creates a carcinogenetic environment challenging for direct measurement and prediction due to no epidemiology of patients with cancer in space (Barcellos-Hoff et al. 2015). Space radiation is deleterious for cognitive and behavioral functions (Kiffer et al. 2018) and for the central nervous and cardiovascular systems (McBeth and Borak 2018). Currently, applied countermeasures based on shielding and drugs are not sufficient (Cerri et al. 2016). One solution worth consideration is torpor of the crew, discussed by Margaret Boone Rappaport and Christopher Corbally in this volume. Torpid animals are less sensitive to damage of tissue caused by space radiation (Griko and Regan 2018). Humans also could be induced into a synthetic torpor to protect their health but also to reduce mass of the spacecraft and the stored resources, as well as about 50% of the energy consumption in spacecraft. Hibernation is a good tool to cope with psychological stress and to reduce aging (Cerri et al. 2016). But the study of human torpor during spaceflight should be connected with the development of countermeasures for the hazardous impact of microgravity. The basic method to cope with this challenge is regular physical exercise. Spacecraft could be equipped with a system used to stimulate muscles of the torpid crew (Cerri et al. 2016). Braddock (2018) is one of the advocates of application of torpor to the space crew, which will take into account challenges of the simultaneous muscle atrophy caused by microgravity. The case of torpor shows that one of the most technologically achievable anti-radiation countermeasures is still beyond currently applied solutions. It may be expected that its application will be challenging not only technologically, but also from an ethical point of view.

Space radiation is probably the most dangerous factor in space. While Solar Particle Events (SPEs) occur seasonally, Galactic Cosmic Rays (GCRs) are chronic. The biggest risk associated with SPE is their short, but highly intensive emissions occurring usually every few years. Some of them registered in 1956, 1960 and 1972 have exceeded protective capabilities of spacesuits (Townsend et al. 2018, p. 34). The biggest challenge for the safety of astronauts in space are deadly radiation doses.

14.6.2 Microgravity

The last hazardous factor in space worth mentioning is altered gravity. While space radiation is the biggest hazard for human health in space, living in a microgravitational environment is also challenging for human health adapted to earthly 1 g gravitation. The lack of a gravitational load causes many deleterious effects. One of them is bone loss estimated to be ten times higher than in osteoporosis. A six-month mission on the ISS causes 10% of loss in total mass of bone. Countermeasures for astronauts include diet, drugs, and physical exercises (Ohshima 2012).

Nick Kanas and Dietrich Manzey emphasize the challenge of rehabilitation of astronauts after de-conditioning during spaceflight. While after previous space missions astronauts always come back to Earth, in the case of a mission to Mars astronauts will be obliged to land on Mars—an environment of altered gravity—without recovery support available on Earth after spaceflights (Kanas and Manzey 2008, p. 26).

14.6.3 Psychological Challenges

Space psychology is likely to be in the field of human space sciences, which—in contrary to medical fields working with hazards of radiation and microgravity—is deeply and broadly prepared for possibly all challenges and threats. One of the advantages of space psychology over medical space sciences is that crew selection may be conducted effectively and may really reduce the risk of serious psychological or psychiatric disorders in space. The pre-launch crew selection is an important tool in coping with possible psychological challenges, in contrast to physiological threats. Human physiological variation in, for instance, immune system or genotype is too narrow and too insufficient to offer humans who would be resistant to space radiation or the deleterious impact of the microgravitational environment. In the field of prevention, space psychology has been successful in the current history of orbital and lunar flights—there was no manic-depressive disorder nor schizophrenia. The program of psychological selection including, among others, health history of family of candidates, is effective (Kanas et al. 2009). During human spaceflights, there have been reported no serious mental disorders, killing, suicide, or physical aggression. But it is worth keeping in mind that the Antarctic winter-over expeditions—considered as one of the best terrestrial analogs for Mars missions—show an observed 1–5% rate of psychiatric disorders (Kanas and Manzey 2008, p. 225). While there is no super-human on Earth who could be resistant to the hazardous impact of microgravity and/or space radiation, there might be selected psychological super-humans who are psychologically suited better than others to live in the space environment. Space psychology may apply particular procedures not available for space physiology. These procedures include the mentioned pre-launch crew selection according to such criteria like gender, size of the group, or cultural specificity. While human psychology depends on physiology, psychological hazards in space might be reduced more effectively and independently than physiological challenges.

Some stressors are universal for all kinds of spaceflights. They include social conflicts and discomfort within the crew, and between crew and mission control on ground, anxiety, workload, or sleep deprivation (Kanas 2015). Some stressors and stresses are universal not only in space but also on Earth (Tachibana et al. 2017). It is important to consider the ethical issues of a human Mars mission because we can widely reach an agreement that such stressors and stresses are ethical risks for humankind. Some psychological inconveniences are produced by physiological deprivation, while others are the domain of dynamics of small isolated groups both in space and on Earth (Kanas et al. 2009, p. 661). The same stressors appear on the ISS and in the future mission to Mars, but it may be expected that their intensity may be stronger due to the longevity of a Mars mission.

New challenges appear in a mission to Mars due to its time and distance from Earth. They include no possibility of fast return to home and the constant high risk of death, monotony, communication delay, the Earth-out-of-view phenomenon (Kanas et al. 2009, p. 661), or permanent threat and anxiety (Kanas and Manzey 2008, p. 223). Psychological and physiological stress in space impairs the immune system (Mehta et al. 2012).

Pre-launch crew selection and training for a mission to Mars may be more challenging than for the previous missions. The current crew on the ISS is not necessarily designed to participate in a mission to Mars (Beven 2012). Exploration missions may require new training techniques than currently applied. Kanas et al. (2009, p. 674) add that an important factor will be an appropriate cultural policy.

It is interesting what astronauts and cosmonauts think about a human mission to Mars. One of the surveys conducted with cosmonauts enumerates such expected risks like the long-term isolation in spacecraft, monotony, communication delay with Earth, and limited resources of water and food (Nechaev et al. 2007, p. 352).

To sum up, no one is able to predict and to fully prepare human astronauts psychologically for a mission to Mars. Mission planners and psychologists expect various reactions, but they cannot control nor predict them. It is worth keeping in mind that physiological deprivation during the 5–6-month journey to Mars will affect the psyche and cognitive capabilities. From an ethical point of view, this mission will be a psychological—but also physiological—vivisection.

14.7 Technological and Ethical Challenges of Advancement in Space Missions

Interplanetary missions—because of their duration and distance from Earth—open space for the new technological requirements, which are not needed for the current orbital missions, or even for a human mission to the Moon. These two parameters in a Mars mission introduce two kinds of technological challenges. One of them is connected with the fact that no delivery in reasonable periods of time will be possible between Earth and Mars. The launch window for Mars appears every 26 months. For this reason, spacecraft should be equipped with all required goods both in the sense of their quantity as well as in quality including, for instance, the issue of expiration date not only for food, but for appropriate diet or effective drugs. The next challenge is connected with communication delay and no opportunity to fast return to home, for instance in the case of emergency.

Consider the mentioned storage of food in space. Like some studies show, the biggest challenge is not the impact of space radiation, but long-term storage (Schroeder 2018). Other technological challenges traditionally mentioned in the context of long-term space missions include life support systems, in situ resources utilization, or possibilities of plant cultivation or production of propellant.

Let's focus on an issue that is challenging not only technologically, but also ethically—the idea of human enhancement in space. This challenge is discussed in more detail elsewhere Szocik et al. (2019a in press, in press b; Szocik and Tachibana 2019). This idea—which today for some scientists may sound more like science fiction fantasy—appears as a theoretical proposal and as a consequence of the discussed above differences between human and robotic missions. If human astronauts for various reasons are still preferred also for interplanetary missions, then the broad set of physiological and psychological challenges remains. As was mentioned, currently applied countermeasures are insufficient. Human enhancement is a broad term, which includes various extents of modification of human physiology, cognition, and psychology.

This issue is partially connected with human phenotypic variability. While this variability is limited by adaptive space on Earth (Gyngell 2012), some candidates for astronauts may be equipped with alleles of genes that predispose them better to realize some tasks. For instance, the best athletes possess alleles of the ACE gene, which increase performance and endurance (Irschick and Higham 2016, p. 223). One of the methods of enhancement is an application of testosterone. Extra testosterone increases strength and mass of muscles—mostly when connected with the high-protein diet—and improves repairing of the damaged cells. Ones of consequences of this implementation may be higher ability to risky and aggressive behaviors (Irschick and Higham 2016, pp. 227, 229). As Irschick and Higham (2016, p. 230) note, there are genetic and morphological differences between human athletes and nonathletes.

With regard to space missions, human enhancement could consist of gene editing. It is no doubt that this procedure is controversial and it seems doubtful that public space agencies would decide to consider it for the purpose of the pre-launch crew. Private space investors probably would be more prone to apply controversial solutions, but they will be in any case limited by public law. There are at work at least two scenarios for possible progress in human enhancement in space. One scenario assumes that—due to the ethical issues—no one decides to enhance astronauts before the first mission to Mars. If hazardous medical changes of the first Mars astronauts after their coming back to Earth are too serious, then mission planners—if they decide to continue the human interplanetary program—might start to consider human enhancement. It is worth keeping in mind that a decision on the application of human enhancement in space will be correlated with parallel progress in alternative countermeasures. No one is able to exclude such a possibility. But if progress in science and technology still is insufficient to produce more effective countermeasures, then advocates of a human mission to Mars may be prone to consider human enhancement as a last option.

The second possible scenario for application of human enhancement in space may be as follows. Mission planners may decide to enhance human astronauts before the first mission to Mars. But such decision would require strong incentives from outside of the space industry. In this scenario, progress in other fields—both technological and ethical—may affect ethical attitudes of space communities. For instance, gene editing may become more or less standard practice in some fields of the human life like medicine or the military. Then, the space missions may be the next field in which human enhancement could be acceptable. It may be more possible if human interplanetary missions will be militarized, conducted in the militarized order and excluded from civil rules.

14.8 Conclusions

The main aim of this chapter was to enumerate the main challenges for a human mission to Mars and for human space settlement. The basic challenge is connected with financial opportunities. While public budgets might be able to finance at least a short-term mission to Mars, they are obliged to follow the criteria of social justice and social responsibility. This duty is expected to increase in the near future due to growing existential threats. The next troublesome factor is the risk of social and political collapse, which may occur in the next decades. As was suggested in this chapter, private investors do not seem to be a promising alternative. While they may be effective in operations in low Earth orbit, they probably are not able to finance the human interplanetary program. It is worth keeping in mind the trivial fact that private business—in contrast to the public budgets—is based on the criterion of economic profits. It is not clear which and why private companies would pay for conducting a mission to Mars or Mars settlement. However, some recent facts—associated with a flyby trip to the Moon, not with staying on Mars—may be some exemptions to this rule. A Japanese investor, Yūsaku Maezawa, made a contract with Elon Mask for the first Moon trip. He said that he also keeps some tickets for artists to develop art. We can remember that Bill Gates and other many billionaires spend their money for public purposes such as advancing medical and pharmacological study and solving poverty. While these examples show that economic profits may not be the only criterion for private money and private business, expenditure and risk of a commercial, nonprofit trip to Mars seems to be more challenging than the mentioned examples. Financial, political and social reasons create one group of factors of risk. Associated issue is no strong rationale for this effort. As it was suggested in the context of the paradox of technological development, the progress in space technologies expected in the next decades may be counterbalanced by disadvantageous social and political situation.

If the unique rationale is progress in science, then space robots should be treated as rivals for human astronauts.

Last but not least, the hazardous space environment usually is not treated as an obstacle to send humans in space. The hostile space environment motivates scientists to find solutions and countermeasures. But—as it was suggested in this chapter—maybe this is time to treat risks for human health and life during the interplanetary mission as one of obstacles to send humans to Mars. If progress in science and technology is not able to effectively protect human astronauts in space, this fact, connected with factors discussed above, may be challenging for the sense of this mission.

The mentioned issues emphasize complexity of the so-called human factor in a mission to Mars. This big effort is in fact the combined effect of many fields of human life and activities, which apparently are not connected or are not necessarily considered. Like other contributors in this volume show, particular issues not directly connected with the space program, like ethical issues connected with the presence of life on Mars or its absence, psychological permanent discomfort on Mars and the feeling of senselessness, or the long-term, multigenerational responsibility for the human space settlement, should be taken into consideration by mission planners in the very long-term perspective.

Acknowledgements Many thanks to Jacob Haqq-Misra and Koji Tachibana for their useful comments.

References

Barcellos-Hoff, M. H., Blakely, E. A., Burma, S., Fornace, A. J., Jr., Gerson, S., Hlatky, L., et al. (2015). Concepts and challenges in cancer risk prediction for the space radiation environment. *Life Sciences in Space Research, 6,* 92–103.

Beven, G. (2012). NASA's behavioral health support for International Space Station (ISS) Missions. Cleveland Clinic Department of Psychiatry and Psychology Grand Rounds. September 13, 2012.

Braddock, M. (2018). Next steps in space travel and colonization: Terraforming, ectogenesis, nano spacecraft and avatars. *Significances of Bioengineering & Bioscience* 2(4). https://doi.org/10.31031/sbb.2018.02.000541.

Cerri, M., Tinganelli, W., Negrini, M., Helm, A., Scifoni, E., Tommasino, F., et al. (2016). Hibernation for space travel: Impact on radioprotection. *Life Sciences in Space Research, 11,* 1–9.

Cohen, M. M., & Haeuplik-Meusburger S. (2015). What do we give up and leave behind? In *45th International Conference on Environmental Systems*, 12–16 July 2015, Bellevue, Washington.

Crawford, I. A. (2012). Dispelling the myth of robotic efficiency: Why human space exploration will tell us more about the Solar System than will robotic exploration alone. *Astronomy and Geophysics 53*, 2.22–2.26.

Do, S., Owens, A., Ho, K., Schreiner, S., & de Weck, O. (2016). An independent assessment of the technical feasibility of the Mars One mission plan—Updated analysis. *Acta Astronautica, 120,* 192–228.

Griko, Y., & Regan, M. D. (2018). Synthetic torpor: A method for safely and practically transporting experimental animals aboard spaceflight missions to deep space. *Life Sciences in Space Research, 16,* 101–107.

Gyngell, Ch. (2012). Enhancing the species: Genetic engineering technologies and human persistence. *Philosophy & Technology, 25,* 495–512.
Irschick, D. J., & Higham, T. E. (2016). *Animal athletes: An ecological and evolutionary approach.* Oxford and New York: Oxford University Press.
Jakosky, B. M., & Edwards, Ch S. (2018). Inventory of CO_2 available for terraforming Mars. *Nature Astronomy, 2,* 634–639.
Kanas, N. (2015). Psychology in deep space. *The Psychologist, 28,* 804–807.
Kanas, N., & Manzey D. (2008). *Space psychology and psychiatry*. Cham: Springer.
Kanas, N., et al. (2009). Psychology and culture during long-duration space missions. *Acta Astronautica, 64*(7–8), 659–677.
Kiffer, F., et al. (2018). Late effects of ^{1}H irradiation on hippocampal physiology. *Life Sciences in Space Research, 17,* 51–62.
Lee, R. B. (1985). Models of human colonization:!Kung San, Greeks and Vikings. In E. Jones, & B. Finney (Eds.), *Interstellar migration and the human experience* (pp. 180–195). University of California Press.
Lester, D. F., Hodges, K. V., & Anderson, R. C. (2017). Exploration telepresence: A strategy for optimizing scientific research at remote space destinations. *Science Robotics 2,* eaan4383.
McBeth, R. A., & Borak, T. B. (2018). Spatial resolution requirements for active radiation detectors used beyond low earth orbit. *Life Sciences in Space Research, 18,* 52–63.
McKay, Ch P, Toon, O. B., & Kasting, J. F. (1991). Making Mars habitable. *Nature, 352,* 489–496.
McKay, Ch P. (2009). Planetary ecosynthesis on Mars: Restoration ecology and environmental ethics. In C. M. Bertka (Ed.), *Exploring the origin, extent, and future of life: Philosophical ethical and theological perspectives* (pp. 245–260). Cambridge: Cambridge University Press.
Mehta, S. K., Pierson, D. L., & Ott, C. M. (2012). Early detection of immune changes prevents painful shingles in astronauts and Earth-bound patients. In *International Space Station. Benefits for Humanity*.
NASA. *NASA Langley Research Center's contributions to the Apollo program.* https://www.nasa.gov/centers/langley/news/factsheets/Apollo.html.
Nechaev, A. P., Polyakov, V. V., & Morukov, B. V. (2007). Martian manned mission: What cosmonauts think about this. *Acta Astronautica, 60,* 351–353.
Ohnishi, T., Takahashi, A., & Ohnishi, K. (2002). Studies about space radiation promote new fields in radiation biology. *Journal of Radiation Research, 43*(SUPPL), S7–S12.
Ohshima, H. (2012). Preventing bone loss in space flight with Prophylactic Use of Bisphosphonate: Health Promotion of the Elderly by Space Medicine Technologies. In *International Space Station. Benefits for Humanity*. https://www.nasa.gov/mission_pages/station/research/benefits/bone_loss.html.
Orosei, R., et al. (2018). Radar evidence of subglacial liquid water on Mars. *Science, 25*(July). https://doi.org/10.1126/science.aar7268,aar7268.
Rovetto, R. J. (2013). The essential role of human spaceflight. *Space Policy, 29*(4), 225–228.
Rovetto, R. J. (2016). Defending spaceflight—The echoes of Apollo. *Space Policy, 38,* 68–78.
Salotti, J.-M., & Suhir, E. (2014). Manned missions to Mars: Minimizing risks of failure. *Acta Astronautica, 93,* 148–161.
Schroeder, R. (2018). Microgels for long-term storage of vitamins for extended spaceflight. *Life Sciences in Space Research, 16,* 26–37.
Shelhamer, M. (2017). Why send humans into space? Science and non-science motivations for human space flight. *Space Policy, 42,* 37–40.
Slakey, F., & Spudis, P. D. (2008). Robots vs. Humans: Who should explore space? *Scientific American.* https://www.scientificamerican.com/article/robots-vs-humans-who-should-explore/
Smith, C. M., & Davies, E. T. (2012). *Emigrating beyond Earth: Human adaptation and space colonization*. New York: Springer.
Szocik, K. (2019a). Should and could humans go to Mars? Yes, but not now and not in the near future. *Futures 105,* 54–66.

Szocik, K. (Ed.). (2019b). *Human enhancements in Lunar, Martian, and future missions to the outer planets.*

Szocik, K., & Tachibana, K. (2019). Human enhancement or "dangerous" AI? Ethical consequences of advanced progress in human and uncrewed space program (in press).

Szocik, K., Campa, R., Rappaport, M. B., & Corbally, Ch. (In press a). *Changing the paradigm on human enhancements. The special case of modifications to counter bone loss for manned Mars missions.*

Szocik, K., Wójtowicz, T., Rappaport, M. B., & Corbally, Ch. (In press b) *Mission to Mars: A challenge for the ethics of human health and biology.*

Tachibana, K., Tachibana, S., & Inoue, N. (2017). From outer space to Earth—The social significance of isolated and confined environment research in human space exploration. *Acta Astronautica, 140,* 273–283.

Townsend, L. W., et al. (2018). Solar particle event storm shelter requirements for missions beyond low Earth orbit. *Life Sciences in Space Research, 17,* 32–39.

Valentine, D. (2017). Gravity fixes: Habituating to the human on Mars and Island Three. *Hau: Journal of Ethnographic Theory, 7*(3), 185–209.

Weinberg, S. (2013). Response: Against manned space flight programs. *Space Policy, 29*(4), 229–230.

Weintraub, D. A. (2018). *Life on Mars. What to know before we go.* Princeton University Press: Princeton and Oxford.

Index

K. Szocik (ed.), *The Human Factor in a Mission to Mars*,
Space and Society, https://doi.org/10.1007/978-3-030-02059-0

Printed by Printforce, the Netherlands